マリタイムカレッジシリーズ

Let's Enjoy Maritime English

商船高専キャリア教育研究会 編
MAAP 協力

KAIBUNDO

著 者

Jane D. Magallon (MAAP)

フィリピンのセブ島の出身で英語，セブ語，タガログ語を話す。セブノーマル大学教養学部にて教養（英語専攻）学士号，フィリピンノーマル大学大学院英語学専攻にて教養（英語教育専攻）修士号を取得。現在MAAPで海事英語教員を13年務めている。MAAPの練習船Kapitan Felix Oca（航海訓練所の旧青雲丸）による乗船実習などにも積極的に参加し，海事英語教育の向上に努めている。最近は同校の海事英語教育システムの評価を担当。フィリピン言語研究学会，IMLA-IMECの世話役も務めている。

Author

She comes from Cebu island in the Philippines and speaks English, Cebuano and Tagalog. She got a bachelor's degree in Arts (major in English) in Cebu Normal University, and has her master's degree in English Language Arts in the Philippines Normal University. She has been teaching Maritime English course and other English courses such as grammar and composition, technical writing and speech communication in the MAAP for 13 years. She is also developing an assessment program in the MAAP at present. She has been on-board the training ship at Kapitan Felix Oca (the previous Seiun Maru 1 of NIST Japan). She is a member of the Philippines Linguistics Society and a Steering Committee member of the IMLA-IMEC, an international professional organization of the IMO Maritime English.

柳沢　修実（弓削商船高等専門学校商船学科）
村上　知弘（弓削商船高等専門学校商船学科）
坂内　宏行（弓削商船高等専門学校総合教育科英語）
清水　聖治（大島商船高等専門学校商船学科）
朴　　鍾徳（大島商船高等専門学校商船学科）
保前　友高（富山高等専門学校商船学科）
山本桂一郎（富山高等専門学校商船学科）
今井　康之（鳥羽商船高等専門学校商船学科）
窪田　祥郎（鳥羽商船高等専門学校商船学科）
木下　恵介（広島商船高等専門学校商船学科）

イラスト

五井　和貴（弓削商船高等専門学校専攻科学生）
前畑　航平（大島商船高等専門学校商船学科）

……はじめに……

かつて全盛を誇っていた日本人船員も日本の経済発展やグローバル化に伴いその数を大幅に減らしてきました。これに代わって，フィリピン，インド，インドネシア，クロアチアなどの外国人船員が日本の外航船社の船員の大多数を占めるようになりました。

この厳しい状況を打破すべく，我々，商船学科を持つ5高専は産学連携助成事業「ALL SHOSEN 学び改善プロジェクト」「海事人材育成プロジェクト」を推進し，有能な日本人の外航船員，船舶管理職員，海外駐在員の育成を鋭意行ってきました。外航船員には高いコミュニケーション能力が必要とされます。そこで，このプロジェクトの一つとして英語力の育成を行っています。また，英語を母国語としない人同士の英語コミュニケーション能力の育成にも力を入れています。

多くの有能な外航船員を輩出するフィリピンの商船学校Maritime Academy of Asia and the Pacific（MAAP）の全面的協力を得て，商船学科を持つ5高専で学生向けの「MAAP海事英語セミナー」と教員向けの「MAAP海事英語教授法セミナー」を実施してきました。この度，これらのセミナーの内容を吟味，編集し，本書を出版するに至りました。

本書ではマリタイムカレッジシリーズ『Navigating English』などで基礎的な英語を学んだ学生が最後の仕上げとして，商船の現場で即時に使える海事英語を学習することを第一の目標とします。ゲームやロールプレイをふんだんに取り入れ，学生が主体的に楽しく学習できることを第二の目標とします。英語を母国語としない人たちと英語を用いて円滑なコミュニケーションを行えるようになることを第三の目標とします。

今日の円安傾向，経済発展に伴う外国人船員の給料水準の上昇，日本の生活インフラを支える国家安全保障上の観点などから，日本の外航船社による日本人船員採用の機運が高まっており，高専の商船学科の学生にとっては就職を勝ち取るチャンスが到来しています。ぜひ，本書を使って世界水準の海事英語を身につけ，このチャンスを我がものにしてください。

謝　辞

「海事人材育成プロジェクト」の一環として本書を出版することができました。全日本海員組合，国際船員労務協会（IMMAJ），Maritime Academy of Asia and the Pacificをはじめとする関係者のみなさまの御協力に感謝します。

······ Introduction ······

The number of Japanese international ship officers has decreased with the economic development and globalization of Japan. Currently, foreign ship officers from Philippines, India, Indonesia, and Croatia get a majority of the employment on-board international ships.

Five colleges of Japan with a maritime department have been trying to educate international ship officers, ship management officers and overseas representatives through the "All Maritime Learning Improvement Project" and "Kaiji Jinzai Ikusei Project" (KJIP). Since the shipping department students need high English communication skill, the "MAAP Maritime English" seminar for students and "MAAP Maritime English Education" seminar for teachers were held with the cooperation of the Maritime Academy of Asia and the Pacific (MAAP) that educates international ship officers in the Philippines. The seminar contents have been edited and published into this textbook.

This advanced textbook is targeted for the students who have already studied basic maritime English, using the textbook "Navigating English" in Maritime College Series. First, the students can study real English used on their job in the maritime industry with the textbook. Secondly, the students can study actively through many games, pair work, group work and role play. Third, the students can study English communication skills not only with the native English speakers but also among themselves who are nonnative speakers of the English language as they practice speaking in some lesson's activities.

The big change came due to weaker Japanese Yen, increment of ship officers' salary level, and the national security. Japanese shipping companies have been trying to get Japanese ship officers for their ocean-going ships. The authors hope that students using this textbook will get a world-class standard English communication skill in order to get a job in the shipping industry which is one of the most important infrastructures in Japan.

Acknowledgments

The publication of this textbook is sponsored by KJIP. We are grateful to All Japan Seamen's Union (JSU), International Mariners Management Association of Japan (IMMAJ), and Maritime Academy of Asia and the Pacific (MAAP).

Contents

Introduction 4

Lesson 1 Vocabulary

Learning Objectives 7
1.1 Parts of a Ship N E 9
1.2 Basic Safety Training N E 10
1.3 Basic Navigation N 11
1.4 Navigational Equipment N 12
1.5 Winch N 13
1.6 Anchoring N 14
1.7 Weather Broadcasting N 15
1.8 Coastal Navigation (Chart) N 16
1.9 Maritime Buoyage System N 17
1.10 VTS Communication N 18
1.11 Meteorological Condition N 19
1.12 Radar N 20
1.13 Navigation Buoyage System N 21
1.14 Main Engine-1 E 22
1.15 Main Engine-2 E 23
1.16 Main Switch Board E 24
1.17 Electric Generator E 25
1.18 Hand Over of Watch E 26
1.19 Electronic Technology E 27
1.20 Engine Cylinder E 28
1.21 Electric Motor E 29
1.22 Multi-stage Centrifugal Pump E 30
1.23 Heat Exchanger E 31
1.24 Refrigeration System E 32
1.25 3-stage Compressor E 33
1.26 Purifier E 34
1.27 Steering Gear E 35
1.28 Marine Fuels E 36
1.29 Marine Valves E 37
1.30 Boiler E 38
1.31 Boiler Operation E 39

Lesson 2 SMCP

Learning Objectives ······ 41
2.1 International Phonetic Alphabet N ······ 43
2.2 Message Markers N ······ 44
2.3 Hand Over of the Watch N ······ 46

Lesson 3 Listening and Speaking

Learning Objectives ······ 53
3.1 Twist Your Tongue and Thinking Exercise N ······ 56
3.2 Hand Over of the Watch E ······ 57
3.3 Fire Fighting N E ······ 58
3.4 Windlass Trouble upon Leaving the Anchorage Area N E ······ 61
3.5 Bridge and Engine Simulation on Pre-departure Procedures N E ······ 62
3.6 Listening Test N E ······ 67
3.7 Basic Work / Trouble Shooting E ······ 68

Lesson 4 Reading and Writing

Learning Objectives ······ 69
4.1 The Points of Hazards N E ······ 71
4.2 What is the Danger? N ······ 74
4.3 What is the Danger? E ······ 75
4.4 Situational Pictures for Conversation N E ······ 77
4.5 Navigation Log Book Writing N ······ 81
4.6 Engineer's Log Book Writing E ······ 83
4.7 Basic Safety N E ······ 85

Answers ······ 92

Maritime English Vocabulary Terms ······ 99

N 航海コース向け
E 機関コース向け
N E 共通

Lesson 1

Vocabulary

Learning Objectives

At the end of the lessons, the students are able to:

a) identify the maritime techincal professional words used in every subject.

b) determine the meaning of the maritime technical words and their functions.

'Speech ··· is given shape by vocabulary and grammar of the language, presented in a train of sound.' (G. Broughton, et.al.)

Each language system co-exists with other systems and should be taught and used with communication skills. No language system exists in isolation.

When studying vocabulary

- Know the meaning and use of the word in context (connotations)
- Be able to pronounce the word on its own and in connected speech
- Understand the grammatical collocations

学習目標

商船の世界には日常ではほとんど用いられない独特な海事英単語が数多くあります。それらを学ぶことが海事英語学習のなかで大きな割合を占めています。

商船学科の学生のみなさんは，すでに校内練習船による航海実習でかなりの数の海事英単語を学習していると思います。船橋，機関制御室，機関室などにある機器などには日本語とともに英語が併記されているので，学習の助けになっているのではないでしょうか。英語そのものが用語になっているものもかなりありますね。

Lesson 1では，いろいろなゲームなどを通して，楽しく海事英単語の学習数を増やしていきましょう。グループをつくってメンバーで話し合いながら学習してもよいでしょう。また，グループ毎にゲームの得点を競い合うのも効果的でしょう。

また，海事英単語には長い商船界の歴史のなかで育まれたものも数多くあります。語彙の歴史的背景を知ることも楽しいと思いますよ。

学習方法

○写真やイラスト（1.1，1.4，1.5，1.12，1.13，1.14，1.16，1.20，1.21，1.22，1.25，1.26，1.30）

写真などのなかの番号に対応する海事英単語を選んだり書いたりしよう。

○連想ゲーム（1.6，1.7，1.15，1.17）

ページの中心に書いてある海事英単語から連想される海事英単語を書き入れ，そこからさらに連想される海事英単語を書き入れていきましょう。このようにすると，より多くの単語が覚えられます。また，単語同士の関連性も学習することができます。

○クロスワードパズル（1.9，1.19）

横書き用（Across）のキーワードと縦書き用（Down）のキーワードを，クロスする部分に入るアルファベットが同じになるように，升目に書き入れてパズルを完成させましょう。

○欠落文字探し（1.2，1.8，1.10，1.18，1.29）

問題の海事英単語中で欠けている文字を書き入れたり，間違っている文字を直しましょう。

1.1 Parts of a Ship

N E

1

1 **Look at the picture below and identify the numbered basic parts of a ship. Write the number of the part next to its name.**

① Funnel ______
② Propeller ______
③ Bow ______
④ Bridge ______
⑤ Stern ______
⑥ Railings ______
⑦ After deck ______
⑧ Bulwark ______
⑨ Ensign ______
⑩ Anchor ______
⑪ Hawsehole ______
⑫ Scuttles ______
⑬ Foredeck ______
⑭ Keel ______

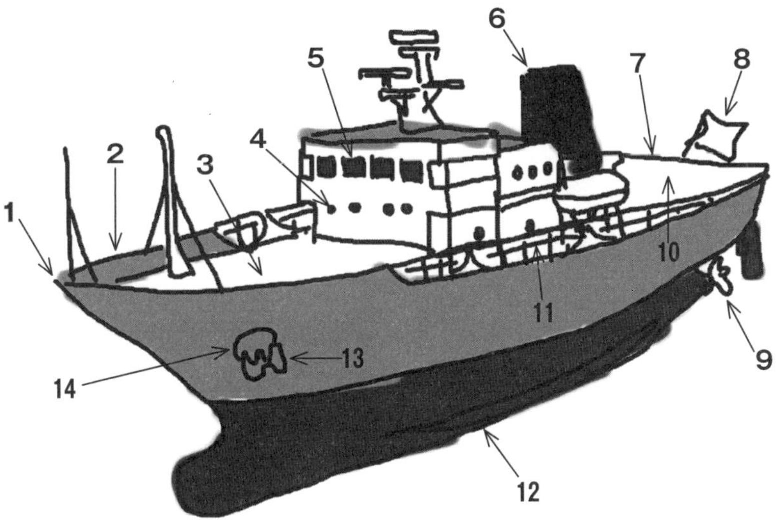

2 **Using the same terms as above, identify the parts described in the following sentences.**

Example : Scuttles : A small opening or hatch.

① ______________________ : The point of a ship that is most forward.

② ______________________ : A weight that is cast overboard to hold a ship fast.

③ ______________________ : The deck towards the stern.

④ ______________________ : A National flag.

⑤ ______________________ : An opening in the bows for a cable.

⑥ ______________________ : A guard around a deck.

⑦ ______________________ : A structure running from bow to stern.

⑧ ______________________ : The part of a ship's side above the deck.

1.2 Basic Safety Training

N **E**

Hidden Personal Protective Equipment (PPE) and Life Saving Apparatus (LSA)

Fill in the missing letters to form the 10 words. Clues are given inside the parentheses. On the line next to the Personal Protective Equipment or the Life Saving Apparatus, write the part of speech of the word formed from the given capital letters.

Example : Life b u o y   Lifebuoy ________ (to be called life ring)

① _ _ _ MET   ________ (to be worn in the head)

② _ LOVE _   ________ (to be worn in the hands)

③ COVER _ _ _   ________ (to be worn in the body)

④ SAFE _ _   _ _ _ _ _   ________ (to be worn on the feet)

⑤ LIFEBO _ _   ________ (life-saving watercraft)

⑥ _ _ _ _ _ _ ION S _ _ _   ________ (to be worn in the body to keep warm)

⑦ GO _ _ _ _ _   ________ (to protect the eyes)

⑧ EAR _ _ _ _   ________ (to protect the ears)

⑨ _ _ _ _   _ ASK   ________ (to be worn to protect the face)

⑩ HA _ _ _ _ S   ________ (tied around the body to prevent falling)

1.3 Basic Navigation

N

Work in Pairs

Choose and underline the correct word from a, b or c. Then, check the pronunciation of the correct words with your partner by reading the sentences aloud.

① If the anchor is not clear, it is [ ].

(a) fault
(b) foil
(c) foul

② When the ship moves forward, she moves [ ].

(a) head
(b) ahead
(c) heading

③ We are approaching the fairway [ ].

(a) boy
(b) buoy
(c) bay

④ [ ] course 245 degrees.

(a) Steer
(b) Stear
(c) Shear

⑤ Make the head lines fast to the [ ] on the forecastle.

(a) bits
(b) bites
(c) bitts

⑥ Have you taken the compass [ ] of the buoy?

(a) steering
(b) bearing
(c) shearing

⑦ The [ ] of the ship measured on the main beam is 21 m.

(a) breast
(b) breath
(c) breadth

⑧ The anchor is [ ].

(a) away
(b) aweigh
(c) awake

1.4 Navigational Equipment

Ⓝ

1 Look at the picture below and identify the numbered equipment at the bridge. Write the number of the equipment next to its name.

① Rudder angle indicator ______

② Gyro compass ______

③ Magnet compass ______

④ Radar ______

⑤ ECDIS ______

⑥ AIS ______

⑦ Microphone ______

⑧ Telephone ______

⑨ Telegraph ______

⑩ VHF radio ______

⑪ Thruster controller ______

⑫ Wing angle meter ______

⑬ Binoculars ______

2 Using the same terms as above, identify the parts described in the following sentences.

Example : ECDIS : A computer-based navigation information and electronic chart system.

① ________________ : To send engine orders from bridge to engine control room.

② ________________ : To show heading of the ship using magnet physical properties.

③ ________________ : To identify and locate vessels by exchanging data with other nearby ship.

④ ________________ : A type of compass based on a fast spinning disc and rotation of the earth.

⑤ ________________ : To communicate with bridge and engine control room.

⑥ ________________ : To show wing angle for the controllable pitch propeller.

⑦ ________________ : Telecommunication device through very high frequency radio.

1.5 Winch

Ⓝ

1

1 **Look at the picture below and identify the numbered basic parts of a winch. Write the number of the part next to its name.**

① Motor ______

② Brake pedal ______

③ Drum ______

④ Shore line ______

⑤ Auxiliary drum ______

⑥ Clutch ______

⑦ Frame ______

⑧ Reduction gear ______

⑨ Brake band ______

⑩ Clutch handle ______

2 **Using the same terms as above, identify the parts described in the following sentences.**

Example : Shore line : Line connected to pier.

① ______________________ : To attach and detach shaft.

② ______________________ : Give rotating power to winch.

③ ______________________ : Min structure to maintain parts.

④ ______________________ : To reduce rotational speed and give more torque.

⑤ ______________________ : To coil anchor chain.

1.6 Anchoring

N

Group Work

Study the word spider below and see how much vocabulary you can generate, with reference to the topic of anchoring.

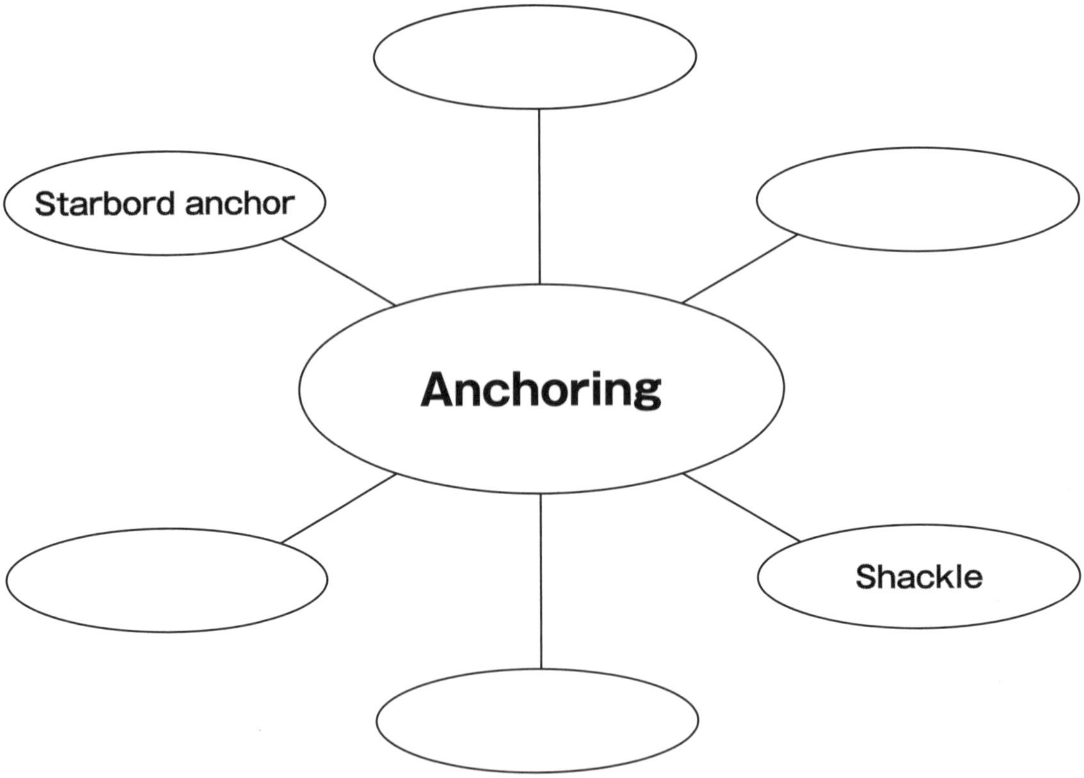

1.7 Weather Broadcasting

(N)

Group Work

Study the word spider below and see how much vocabulary you can generate, with reference to the topic of weather brodcasting.

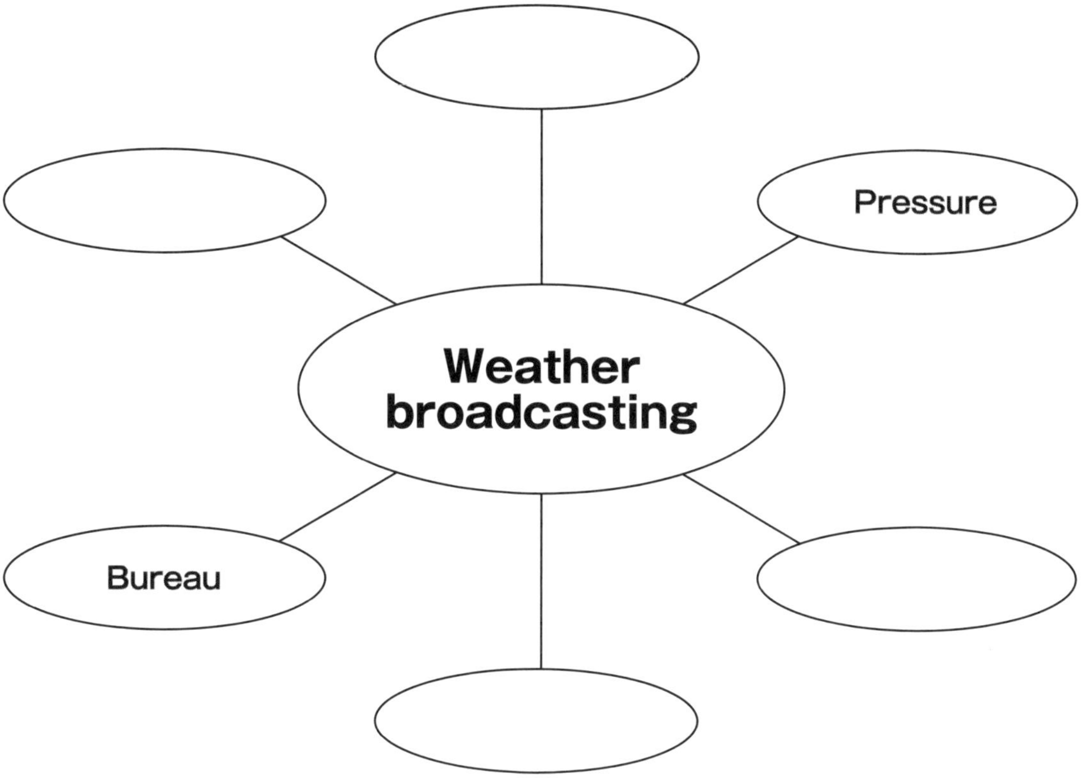

1.8 Coastal Navigation (Chart) Ⓝ

All 26 of the alphabet's letters, one per word, have been deleted. In Example, the **h** is missing from the word – **chart**. Remember that each letter has been deleted only once. Cross out each letter from the list below.

a	b	c	d	e	f	g	~~h~~	i	j	k	l	m
n	o	p	q	r	s	t	u	v	w	x	y	z

Example : <u>chart</u> cart

① ________________ channl

② ________________ reat circle

③ ________________ ltitude

④ ________________ ongitude

⑤ ________________ ompass

⑥ ________________ eviation

⑦ ________________ earing

⑧ ________________ natical

⑨ ________________ mle

⑩ ________________ life acket

⑪ ________________ enith

⑫ ________________ curse

⑬ ________________ magetic

⑭ ________________ gro

⑮ ________________ ais

⑯ ________________ noth

⑰ ________________ assage

⑱ ________________ peed

⑲ ________________ trac

⑳ ________________ airway

㉑ ________________ proection

㉒ ________________ ercator

㉓ ________________ aypoint

㉔ ________________ naigation

㉕ ________________ uarter

1.9 Maritime Buoyage System

Ⓝ

Look at the words listed below. They are wrongly written or spelled as either there is a missing letter or a wrong letter is used. Using the puzzle write the correct spelling of each word.

Across

① Latiral

② Cardnl mark

③ Chanel

④ Seeward

⑤ Obstracion

⑥ Boy

⑦ Conecal

⑧ Navigabl

Down

⑨ Waterlin

⑩ Pilar

⑪ Direcion

⑫ Starbored

⑬ Datom

⑭ Porside

⑮ Spur

1.10 VTS Communication N

Each of the sentences has a common problem – the vowels are gone. The number on top of each word tells how many letters, including the missing vowels, are in the word. Write each sentence on the space provided.

Example : 〔Question〕

3	4	9	5	4	4	3	3
Th	sh	ometmes	nds	elp	fr	th	V .

〔Answer〕 The ship sometimes needs help from the VTS.

①

3	3	4	11	4	3	4
Th	VTS	sks	nfrmtn	frm	th	shp.

____

②

3	6	7	8	2	11	2	7	3	6	2	3	5
Th	vsl	trffc	srvcs	s	stblshd	t	mprv	th	sfty	f	th	shps.

____

③

3	3	7	3	5	2	7	3	7	5
Th	VTS	rdrd	th	shps	t	bsrv	th	frwy	spd.

____

④

4	3	4	7	2	3	9	5	3	3	2	6	2	3	3
Whn	th	shp	rrvs	t	th	rprtng	pnt,	sh	hs	t	rprt	t	th	VTS.

____

⑤

9	3	5	2	3	7	10	6	4	6	2	3	5
bsrvng	th	rls	n	th	trffc	sprtn	schm	dds	sfty	t	ll	shps.

____

1.11 Meteorological Condition N

The words listed below can be found inside the box by connecting the letters. Circle the letters horizontally, vertically or diagonally. Some of the words are written in reverse.

E	N	I	A	R	N	A	M	O	V	E	M	E	N	T
V	I	S	I	B	I	L	I	T	Y	C	J	M	O	R
P	R	E	S	S	U	R	E	A	G	R	A	N	I	O
S	U	I	D	A	R	L	L	O	N	O	K	D	T	P
L	S	E	L	I	M	E	T	E	R	F	B	I	I	I
A	W	I	N	D	I	M	I	L	E	B	A	R	S	C
C	E	V	I	N	S	P	E	E	D	E	C	A	O	A
S	L	G	R	F	T	N	A	C	I	A	K	A	P	L
A	L	P	L	O	O	V	O	E	M	U	I	E	L	A
P	U	T	C	L	O	G	R	W	D	F	N	I	E	O
O	N	J	C	E	A	N	S	U	S	O	G	S	N	I
T	A	Y	M	L	E	I	V	E	E	R	I	N	G	T
C	C	G	E	L	O	C	R	Y	T	T	O	G	O	U
E	D	F	O	R	E	I	M	R	O	T	S	V	E	M
H	Y	D	R	O	L	O	G	I	C	A	L	R	J	N

① Gale
② Storm
③ Tropical
④ Wind
⑤ Backing
⑥ UTC
⑦ Mile
⑧ Cyclone
⑨ Hydrological
⑩ Position
⑪ Hectopascals
⑫ Movement
⑬ Knot
⑭ Raduius
⑮ Visibility
⑯ Sea
⑰ Swell
⑱ Meter
⑲ Icing
⑳ Miles
㉑ Speed
㉒ Force
㉓ Beaufort
㉔ Rain
㉕ Fog
㉖ Veering
㉗ Snow
㉘ Mist
㉙ Pressure

1.12 Radar

Ⓝ

1 Look at the picture below and identify the numbered basic parts of the radar screen. Write the number of the part next to its name.

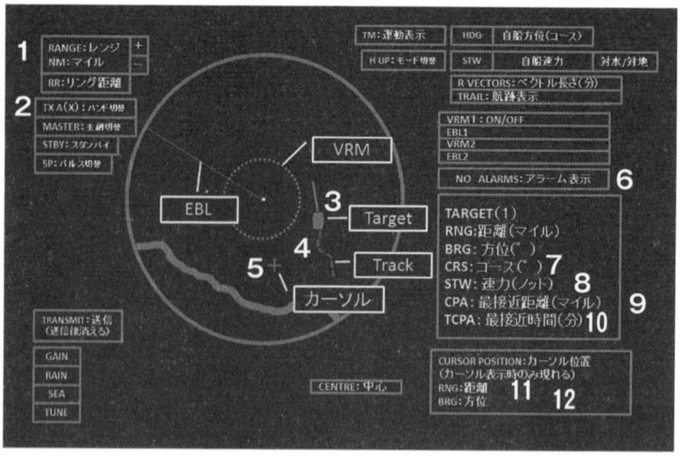

① Range ______

② Band ______

③ Position (Cursor) ______

④ Track ______

⑤ Target ______

⑥ Alarms ______

⑦ RNG (Distance) ______

⑧ BRG (Orientation) ______

⑨ CRS (Vessel course) ______

⑩ STW (Vessel speed) ______

⑪ DCPA (Distance of Closest Point Approach)

______

⑫ TCPA (Time of Closest Point Approach)

______

2 Using the same terms as above, identify the parts described in the following sentences.

① ______________ : Ship which you want to check.

② ______________ : Distance at Closest Point Approach if ship go with same condition.

③ ______________ : Time of Closest Point Approach if ship go with same condition.

④ ______________ : Line where target ship moved.

⑤ ______________ : Frequency band of radar.

1.13 Navigation Buoyage System Ⓝ

Tell what the following buoys mean:

①

____

②

____

③
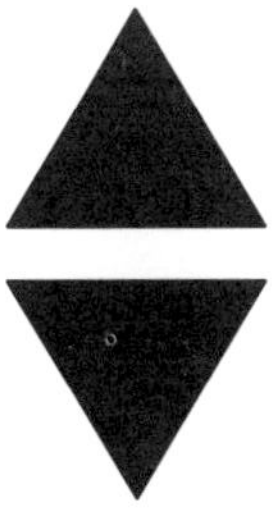
____

④

____

⑤
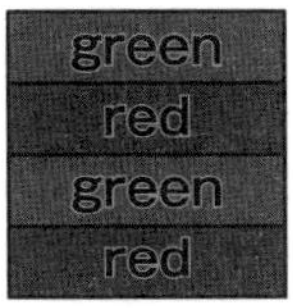

____

⑥

____

⑦
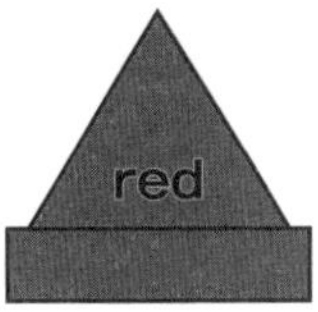

____

⑧

____

⑨

____

1.14 Main Engine-1

E

1 **Look at the picture below and identify the numbered basic parts of the main engine. Write the number of the part next to its name.**

① Governor ______

② Sea water pump ______

③ Clutch ______

④ Bet ______

⑤ Cylinder ______

⑥ Flywheel ______

⑦ FWC
(Fresh Water Cooler) ______

⑧ Cam shaft ______

⑨ Turning gear ______

⑩ Reduction gear ______

⑪ Fuel injection pump ______

⑫ Intermediate shaft ______

⑬ Turbocharger ______

⑭ Crank case ______

2 **Using the same terms as above, identify the parts described in the following sentences.**

Example : Clutch : A mechanism for connecting and disconnecting a vehicle engine from its transmission system.

① ______________________ : A heavy revolving wheel to increase the machine's momentum.

② ______________________ : A device automatically regulating the supply of fuel to an engine.

③ ______________________ : A gearwheels in which the driven shaft rotates more slowly than the driving shaft.

④ ______________________ : A supercharger driven by a turbine powered by the engine's exhaust gases.

⑤ ______________________ : To rotate engine by external motor.

⑥ ______________________ : Transmit rotation from engine to propeller.

⑦ ______________________ : The base of engine.

1.15 Main Engine-2

E

Group Work

Study the word spider below and see how much vocabulary you can generate, with reference to the topic of the main engine.

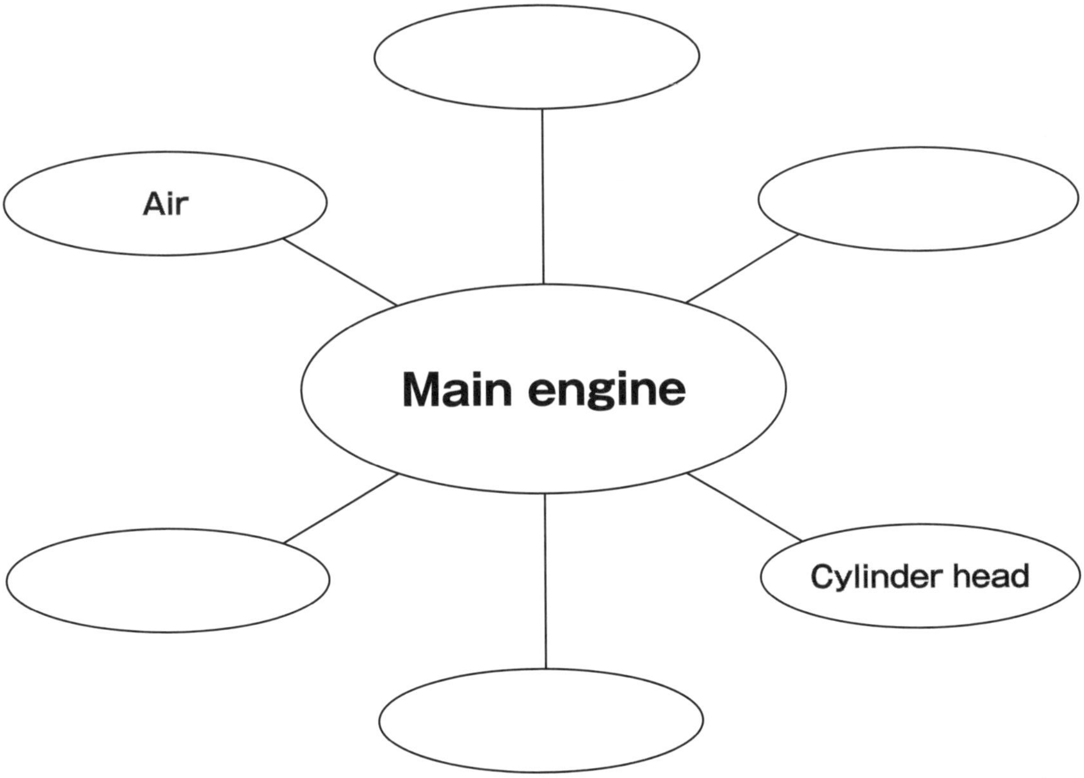

1.16 Main Switch Board

E

1 Look at the picture below and identify the numbered main switch board. Write the number of the part next to its name.

① Starter ______
② Voltmeter ______
③ Synchro scope ______
④ MSB (Main Switch Board) ______
⑤ BUS ______
⑥ Watt meter ______
⑦ ACB (Air Circuit Breaker) ______
⑧ Frequency meter ______
⑨ Shaft generator ______
⑩ Group starter panel ______
⑪ Governor ______
⑫ No fuse breaker ______
⑬ Diesel generator ______
⑭ Amperemeter ______
⑮ Power factor meter ______

2 Using the same terms as above, identify the parts described in the following sentences.

Example : Power factor meter : An instrument for measuring the ratio of the actual electrical power dissipated by an AC circuit to the product of the r.m.s. values of current and voltage.

① ______________________ : An instrument for measuring electric potential in volts.

② ______________________ : An instrument for measuring electric current.

③ ______________________ : An instrument for measuring electric power in watts.

④ ______________________ : An instrument for measuring rate of electric vibration per second.

⑤ ______________________ : A device to regulate rotational speed of generator.

⑥ ______________________ : A distinct set of conductors in parallel.

⑦ ______________________ : An automatic device for stopping the over current to protect electric circuit as a safety.

⑧ ______________________ : An instrument for measuring phase difference in 2 generator.

1.17 Electric Generator

E

1

Group Work

Study the word spider below and see how much vocabulary you can generate, with reference to the topic of electric generator.

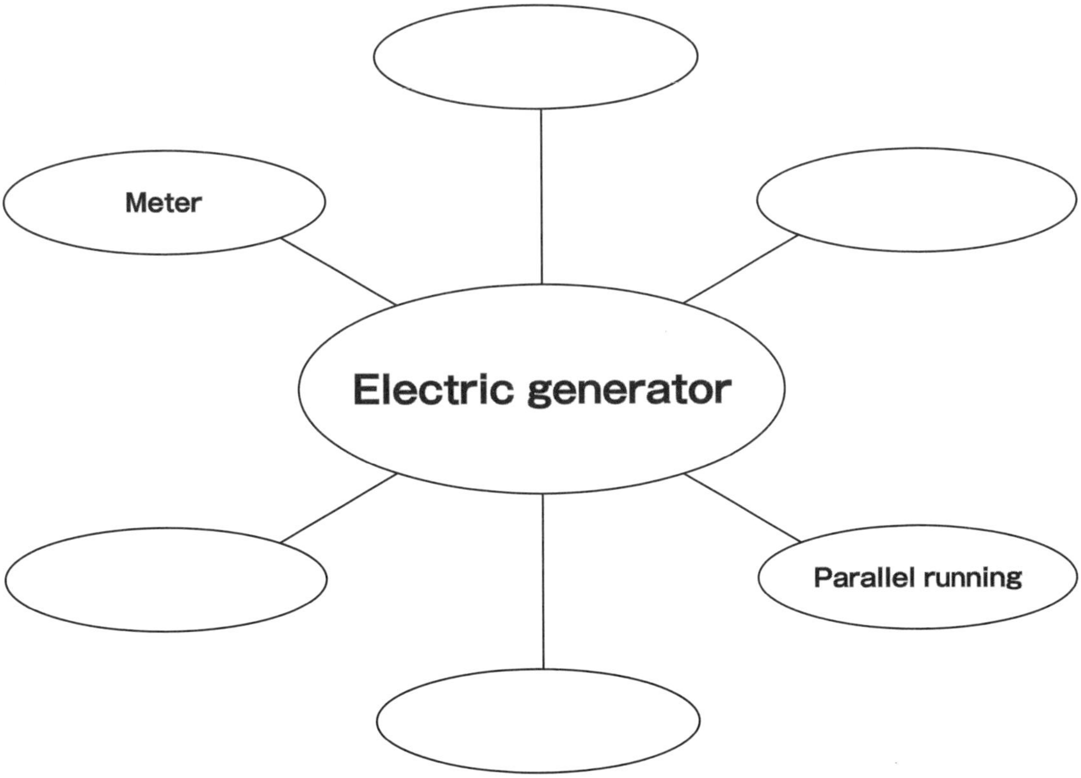

1.18 Hand Over of Watch

E

A letter is missing from each word below. On the first line after the misspelled word, write the letter that is left out and write the correctly spelled word on the long line. If you are correct, your 20 letters will spell the parts of a ship.

① duy ______ ____________________

② min engine ______ ____________________

③ standing order ______ ____________________

④ bellboo ______ ____________________

⑤ machinry ______ ____________________

⑥ tempeature ______ ____________________

⑦ peed ______ ____________________

⑧ ontrol ______ ____________________

⑨ prpeller ______ ____________________

⑩ cylider ______ ____________________

⑪ sucion ______ ____________________

⑫ shft ______ ____________________

⑬ nlet valve ______ ____________________

⑭ ecoomizer ______ ____________________

⑮ gnerator ______ ____________________

⑯ evolution ______ ____________________

⑰ lue oil ______ ____________________

⑱ pmp ______ ____________________

⑲ fue ______ ____________________

⑳ buner ______ ____________________

1.19 Electronic Technology

E

Look at the words listed below. They are wrongly written or spelled as either there is a missing letter or a wrong letter is used. Using the puzzle write the correct spelling of each word.

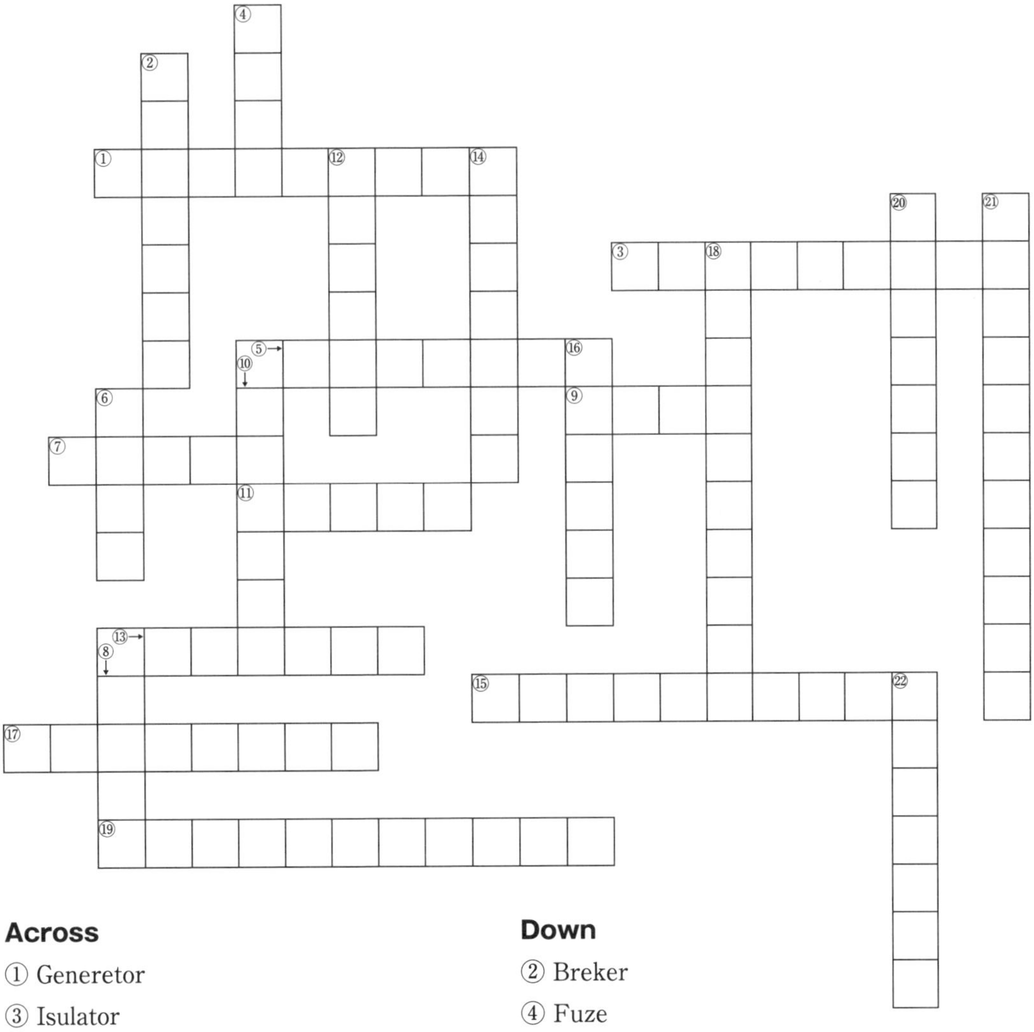

Across

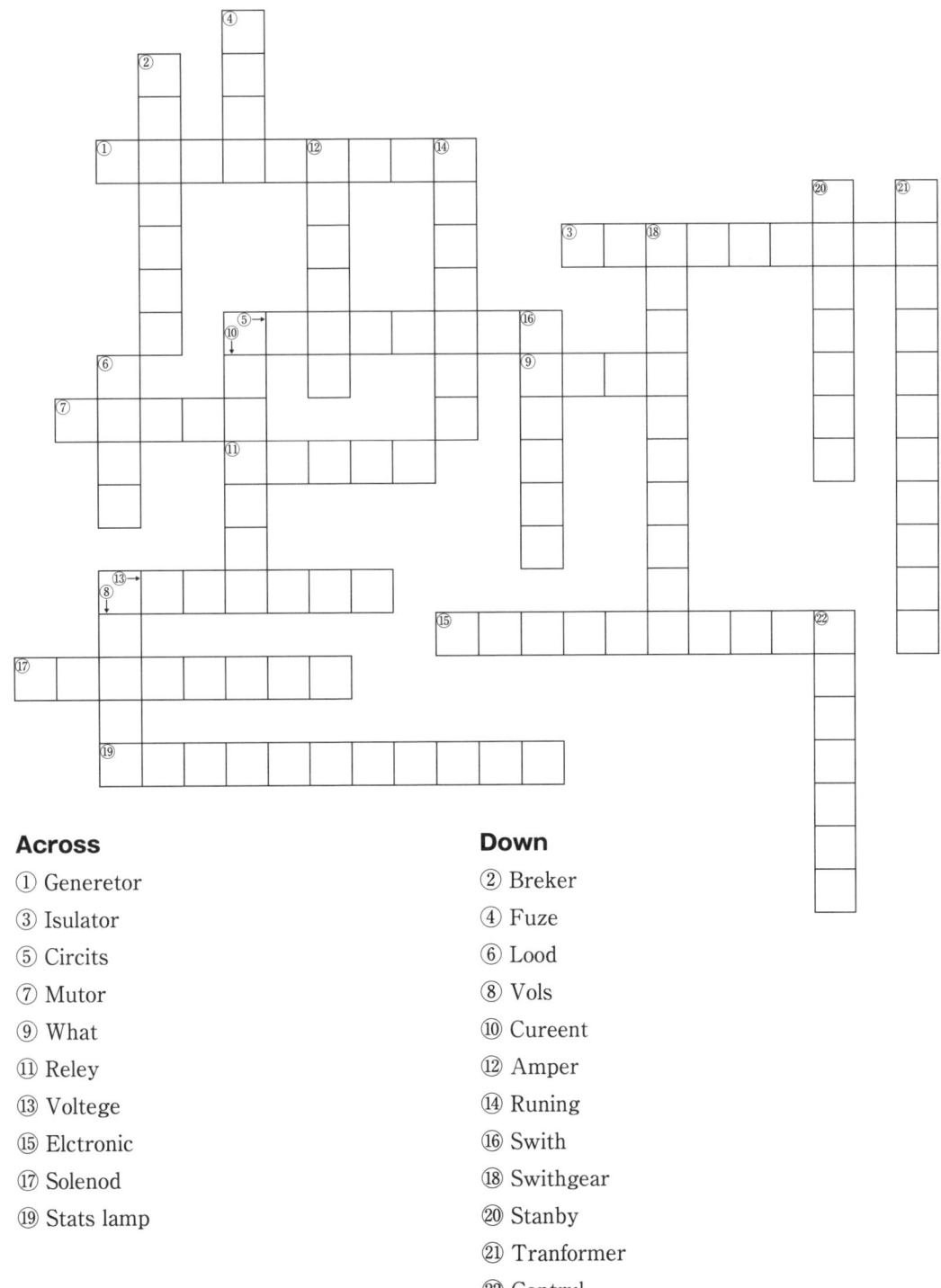

① Generetor
③ Isulator
⑤ Circits
⑦ Mutor
⑨ What
⑪ Reley
⑬ Voltege
⑮ Elctronic
⑰ Solenod
⑲ Stats lamp

Down

② Breker
④ Fuze
⑥ Lood
⑧ Vols
⑩ Cureent
⑫ Amper
⑭ Runing
⑯ Swith
⑱ Swithgear
⑳ Stanby
㉑ Tranformer
㉒ Contrul

1.20 Engine Cylinder

E

1 **Look at the picture below and identify the numbered basic parts of the engine cylinder. Write the number of the part next to its name.**

① Piston ________

② Manifold ________

③ Cylinder ________

④ Crank ________

⑤ Exhaust valve ________

⑥ Lubricating oil ________

⑦ Cam ________

⑧ Piston ring ________

⑨ Fuel injection valve ________

⑩ Cylinder head ________

⑪ Rod ________

⑫ Crank shaft ________

⑬ Inlet valve ________

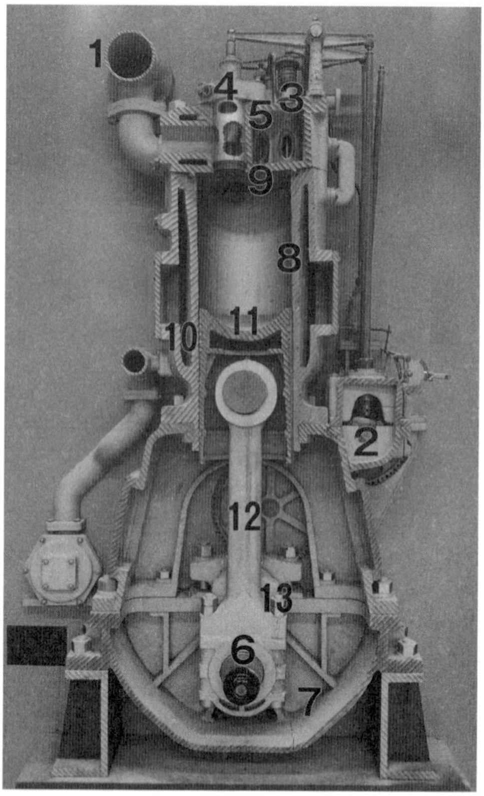

2 **Using the same terms as above, identify the parts described in the following sentences.**

Example : Piston : Cylinder fitting within a tube which moves up and down against a gas.

① ____________________ : Shaft bent out at right angles for converting reciprocal to circular motion.

② ____________________ : For operating the valves.

③ ____________________ : A ring sealing the gap between the piston and the cylinder.

④ ____________________ : A pipe branching into several openings.

⑤ ____________________ : A thin straight bar.

⑥ ____________________ : A solid geometric figure with straight parallel sides and a circular.

⑦ ____________________ : A valve for waste gases expelled from an engine.

1.21 Electric Motor

E

1

1 **Look at the picture below and identify the numbered basic parts of the electric motor. Write the number of the part next to its name.**

① Three-phase AC ________

② Shaft ________

③ Rotor ________

④ Magnetic field ________

⑤ Torque ________

⑥ Terminal ________

⑦ Electromotive force ________

⑧ Bearing bracket ________

⑨ Electric current ________

⑩ Bet ________

⑪ Coil ________

2 **Using the same terms as above, identify the parts described in the following sentences.**

① ____________________ : A region around a magnetic material within which the force of magnetism acts.

② ____________________ : A connector for rotating object.

③ ____________________ : A current of electric charge.

④ ____________________ : A twisting force that tends to cause rotation.

⑤ ____________________ : A length of wire wound in a spiral.

1.22 Multi-stage Centrifugal Pump E

Arrange the letters to form the word or words of the parts of a centrifugal pump.

① ______________________ L D E I R Y V E

② ______________________ E P M I L E L S R

③ ______________________ H A S T F

④ ______________________ C U N I O S T

⑤ ______________________ R A I N E V T V E V A L

⑥ ______________________ N I G C U O P L

⑦ ______________________ E R A B N G I S A C E

⑧ ______________________ M C E A H N I L A C L E S A

⑨ ______________________ S A T H F Y K E

⑩ ______________________ M O V E L U S A C E

1.23 Heat Exchanger

E

Fill in the gaps in the sentences using the words inside the box. All the words are used only once.

Steam	Fuel oil	Fresh water	Three
Electric	Seawater	Plate	Fresh water

Cooling and heating system

① The two types of heat exchangers are the ________________-type and the shell and tube-type.

② There are single, two or ________________ pass in the shell and tube-type of heat exchanger.

③ In low temperature cooler, seawater is used to cool ________________.

④ In high temperature cooler, fresh water is used to cool ________________.

⑤ In surface cooler/condenser, seawater or fresh water is used to cool ________________.

⑥ ________________ is used in cooling the jacket cooling F.W.

⑦ In heating the ________________, the steam or thermal oil or electric can be used.

⑧ Steam or thermal oil or ________________ is used in heating the jacket cooling F.W.

Plate-type cooler or heater

Shell and tube-type cooler or heater

1.24 Refrigeration System

E

Tell where the part of the refrigeration system is. Write the number on the space.

① Where is the condenser? ________

② Where is the compressor? ________

③ Where is the control panel? ________

④ Where is the level gauge? ________

⑤ Where is the pressure gauge? ________

⑥ Where is the cooling inlet pipe? ________

⑦ Where is the refrigerant line? ________

⑧ Where is the insulation? ________

1.25 3-stage Compressor

E

1

Tell the parts of the 3-stage compressor using the given clues and by rearranging the letters to form the word.

① ______________________ R O M T O

② ______________________ C M R O S S O P R E B L E T COVER

③ ______________________ I A R COMPRESSOR D E R Y C L N I

④ ______________________ C U S T N O I D S I H C A G E R VALVE

⑤ ______________________ N I T A E K O R T P

⑥ ______________________ E D R L A O／U N D E R O A L

⑦ ______________________ A S F T E Y VALVE

⑧ ______________________ I D S C A H G R E LINE

1.26 Purifier

E

Name the parts of the purifier using the words inside the box. Write the words on the space below.

Distributor	Bowl nut	Main cylinder	Bowl hood
Impeller	Disc	Bowl body	

① ______________________

② ______________________

③ ______________________

④ ______________________

⑤ ______________________

⑥ ______________________

⑦ ______________________

1.27 Steering Gear

E

1

Draw an arrow (→) from the three main parts of the steering gear system to the picture below.

Control equipment | Power unit | Transmission system

Match the main parts of the steering gear in column A with their functions in column B.

A – Parts		B – Function
① Control equipment	________	a. Gives forces to move the rudder
② Power unit	________	b. Rudder movement is accomplished
③ Transmission system	________	c. Sends signal of rudder angle from the bridge

1.28 Marine Fuels

E

The words listed below can be found inside the box by connecting the letters. Circle the letters horizontally, vertically or diagonally. Some of the words are written in reverse.

S	U	P	E	R	C	H	A	R	G	E	R	J	D	V
Y	T	I	S	N	E	D	N	O	I	T	I	N	G	I
V	A	P	O	R	C	O	R	R	O	S	I	O	N	S
K	N	A	T	N	O	I	S	S	E	R	P	M	O	C
E	N	G	I	N	E	M	A	C	H	E	A	T	X	O
N	D	B	L	O	W	B	Y	O	A	C	I	L	I	S
E	I	P	I	P	E	S	E	M	B	G	A	E	D	I
T	E	L	L	O	U	V	N	B	R	R	E	N	E	T
L	S	J	A	R	I	N	R	U	A	E	M	I	D	Y
A	E	I	V	T	V	Y	U	S	S	K	A	L	I	F
H	L	E	I	O	G	O	F	T	I	N	R	A	R	U
P	Y	D	A	C	I	L	L	I	V	U	I	C	I	E
S	D	A	F	U	E	L	U	O	E	B	N	L	J	L
A	L	U	M	I	N	A	S	N	A	H	E	A	V	Y

① Heavy
② Marine
③ Pipe
④ Engine
⑤ Fuel
⑥ Sulfur
⑦ Viscosity
⑧ Alumina
⑨ Silica
⑩ Abrasive
⑪ Combustion
⑫ Asphaltene
⑬ Blowby
⑭ Supercharger
⑮ Ignition
⑯ Alkaline
⑰ Corrosion
⑱ Alloy
⑲ Additive
⑳ Bunker
㉑ Heat
㉒ Compression
㉓ Oxide
㉔ Oil
㉕ Density
㉖ Vapor
㉗ Survey
㉘ Tank
㉙ Diesel

1.29 Marine Valves

E

1

A letter is missing from each word below. On the first line after the misspelled word, write the letter that is left out and write the correctly spelled word on the long line.

① vlves ______ ____________________

② ody ______ ____________________

③ ras ______ ____________________

④ bnnet ______ ____________________

⑤ cast irn ______ ____________________

⑥ tainless steel ______ ____________________

⑦ bow scrw ______ ____________________

⑧ glnd ______ ____________________

⑨ dis ______ ____________________

⑩ valve set ______ ____________________

⑪ tem ______ ____________________

⑫ leeve ______ ____________________

⑬ paking ______ ____________________

⑭ ctuator ______ ____________________

⑮ glbe valve ______ ____________________

⑯ gte valve ______ ____________________

1.30 Boiler

E

Match the pictures with the major parts of a boiler. Use a line to connect the picture and the part.

① 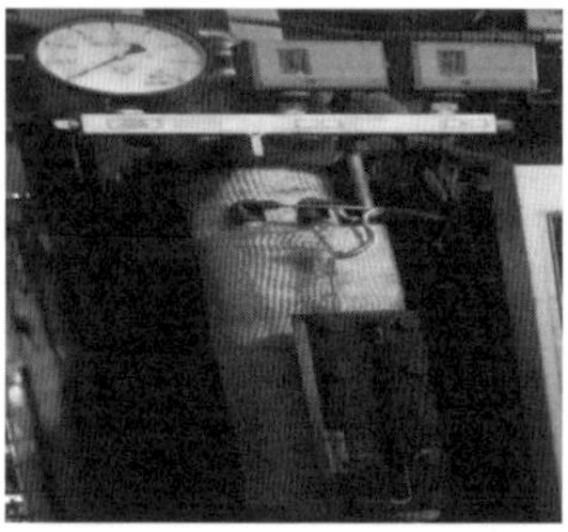

Logic controller

② 

Burner

③ 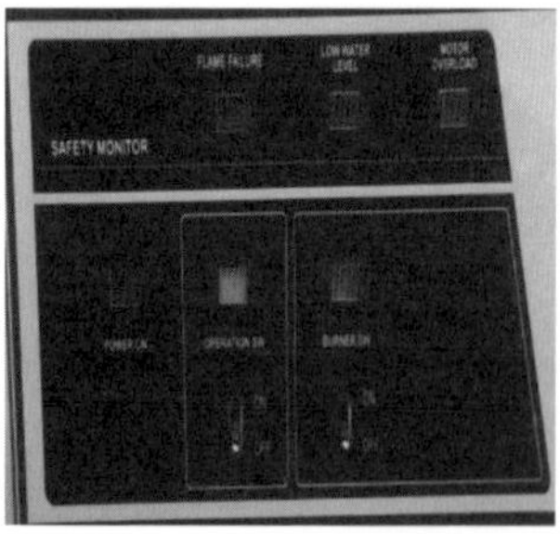

Water level system

④ 

Control system

⑤ 

Safety monitor

1.31 Boiler Operation

E

1

Identify the error in the sentence by writing the number of the word or the phrase and write the correct word.

Example : Make sure (1) all geages (2) are operating (3) correctly (4).

Error	Correct word
(2)	gauges

	Error	Correct word	Sentence
①			Inspect (1) the statuos (2) of the boiler mounting (3).
②			Check (1) the (2) fed water (3) system (4).
③			Check (1) the (2) fewel oil (3) system (4).
④			Switch (1) on the (2) main (3) breker (4).
⑤			Check (1) start interlocks (2) and (3) rectify (4) as necesary (5).
⑥			Put (1) on (2) operetion (3) switch (4).
⑦			Put (1) in (2) burner (3) switch (4).
⑧			Regulate (1) the (2) fuel oil (3) presure (4). (Max. 7 bars, min. 4 bars)
⑨			Check (1) flame propagetion (2), register (3) and smoke (4) from the stack (5).
⑩			Monitor (1) the boiler (2) steam pressure (3) from time to time (4).

Lesson 2

SMCP

Learning Objectives

At the end of the lesson activities, the students are able to:

a) practice in the role play the maritime conversations using the IMO SMCP (Standard Marine Communication Phrases) and identify the maritime technical words.

b) determine the meaning of the maritime technical words and their functions.

○フォネティックアルファベット(2.1)

学習目標

無線機や電話では，音声の完璧な再生は不可能です。必ず歪や雑音が伴います。そこで，情報を正確に伝えるためにフォネティックアルファベットと呼ばれる付加情報を付けます。このフォネティックアルファベットを使えるようにします。

学習方法

①フォネティックアルファベットを覚えましょう。

②自分の名前をフォネティックアルファベットを使用して発音してみましょう。

③航海情報をフォネティックアルファベットを使用して伝えてみましょう。

○メッセージマーカー（2.2）

学習目標

メッセージはInstruction（指示），Information（情報），Intention（意志），Warning（警告），Advice（アドバイス），Request（依頼），Question（質問），Answer（回答）の8つのメッセージマーカーに分類されます。伝える内容を8つのメッセージマーカーに分類し，的確に表現できるようにします。

学習方法

①伝える内容を8つのメッセージマーカーに分類し，的確に表現できるように学習しましょう。

②聞き取った会話を的確に8つのメッセージマーカーに分類できるように学習しましょう。

○ロールプレイ(2.3)

学習目標

船橋における当直交代のシーンを想定して，IMO標準海事コミュニケーション用語・慣用表現を用いてシナリオをつくり，ロールプレイを行い，実務に近い英語会話ができるように学習しましょう。

学習方法

①まずグループに分かれ，船長，一等航海士，甲板員など，ロールプレイ上の一人一人の役割を決めましょう。

②実際に交わす会話をグループ全員で一緒に考え，練習しましょう。

- ほとんどの会話はSMCPによって国際標準化（決まり文句）されています。まずはSMCPを覚えましょう。
- 会話はゆっくり，はっきり発音することを心掛けましょう。
- オーダーを受けた役者は必ずアンサーバックする練習をしましょう。

③実際に練習船の船橋を使いロールプレイを行いましょう。

- 後で復習ができるようにビデオ撮影をしておきましょう。

2.1 International Phonetic Alphabet (N)

Relay the message. Get a partner or pair with anyone and assign Student A who spells out the word using the International Phonetic Alphabet while Student B will receive the massage and write the word.

What is the message? (relay the message)

- Port Toyama
- Cape Tsubaki
- ETA 1530H

IMO International Phonetic Alphabet and numbers is used to spell out words of important information for clear message especially when talking onboard using the VHF radio.

List of Phonetic Alphabet

Letter	Code	Letter	Code	Letter	Code
A	Alfa	J	Juliet	S	Sierra
B	Bravo	K	Kilo	T	Tango
C	Charlie	L	Lima	U	Uniform
D	Delta	M	Mike	V	Victor
E	Echo	N	November	W	Whisky
F	Foxtrot	O	Oscar	X	X-ray
G	Golf	P	Papa	Y	Yankee
H	Hotel	Q	Quebec	Z	Zulu
I	India	R	Romeo		

Numbers are spoken in **single digit** except on rudder commands.

e.g. ETA at Yuge is 1-3-0-0 hours

pos'n lat 1-4 deg 2-2 minutes N; long 1-2-3 deg 1-2 minutes E

Except：Starboard 10 ; Port 15

Try sending the following message below using the International Phonetic Alphabet.

① Search and Rescue finished at 2345 UTC.
② Container in bay 1 row 5 tier 70 is on fire!
③ Sea suction valve number 04 is open.
④ Decrease the pumping rate to 1000 revolutions/bar.
⑤ Check the lashings and securings every 4 hours.
⑥ Winds force Beaufort 3 from Northwest.
⑦ Starboard steer 180 degrees.
⑧ Port 1-5.
⑨ Speed was reduced at 1100 UTC due to bad weather.
⑩ Ballast sounding is 14 cubic meters.

2.2 Message Markers

N

Message marker is placed before the message to prepare the receiver on the information told when talking over the VHF radio.

Message Markers	Example
Question	What is your course? What is your position? How many tugs are required? What is your ETA: Fairway Buoy? What are your intentions?
Answer	My course is 1-3-2 degrees true. My position is: NE of Buoy Number 1.5. I require two tugs. My ETA Fairway Buoy is: time: 1-5-4-5 hours local.
Request	Immediate tug assistance. Please arrange for the berth on arrival. Permission to enter the Fairway. Please send a doctor.
Information	The tanker XEROX is next. My ETA at Outer Pilot Station is ···.
Advice (Strongly recommended, at receiver's option)	(Advise you) Stand by on channel 6-8. Steer course: 2-53-3 degrees true. Anchor in position: bearing: one-two-five degrees true, from Punta Stella, distance two miles.
Instruction (Same as order, command or prohibition)	You must alter course. Stop your engine immediately. Alter course to: new course 1-2-3 true. Push on starboard bow.
Warning	Vessel not under command in ···. Obstruction in the fairway. Tanker aground in position ···. Buoy number: one-five unlit/off position. Pilotage services suspended.
Intention	I intend to alter course to starboard and pass astern of you. I will pass astern of you. I intend to be underway within period: two hours.

1 Put the correct message maker on the blank.

____________ ① You must alter your course to port.

____________ ② Please permit me to enter the restricted zone: reason: steering gear breakdown.

____________ ③ Do not enter the fairway.

____________ ④ Do I have permission to proceed?

____________ ⑤ You are running into danger.

____________ ⑥ No, do not proceed. Wait for further instruction.

____________ ⑦ I will overtake you on your port side.

____________ ⑧ The tide is rising at 0500 hours.

2 Match the message makers and the sentences. Write the letters on the space before the number.

_____ ① Question

_____ ② Answer

_____ ③ Request

_____ ④ Information

_____ ⑤ Advice

_____ ⑥ Instruction

_____ ⑦ Warning

_____ ⑧ Intention

A. Please confirm your deadweight.

B. I will reduce speed.

C. Is buoy Number 1-4 in the correct position?

D. Gale force winds in area ⋯.

E. Wind backing and increasing.

F. Go to berth No. 15.

G. Alter your course to starboard.

H. Negative. Buoy Number 1-4 in not the correct position.

2.3 Hand Over of the Watch

N

Role Play

Using the IMO SMCP below, write a script on how to hand over the watch to the next duty officer. Assign who will be the incoming officer and the outgoing officer.

B1/1 Handing over the watch

B1/1.1 Briefing on position, movements and draft

The officer of the watch should brief the relieving officer on the following:

.1 Position

.1 The present position is

~ latitude …, longitude ….

~ bearing … degrees, distance … cables / nautical miles from/to ….

~ buoy … *(charted name)*.

~ between … and ….

~ way point / reporting point ….

~ ….

.2 The next waypoint / reporting point is ….

.3 ETA at … is … hours UTC.

.4 We are passing / we passed buoy … *(charted name)* on port side / starboard side.

.5 We are approaching buoy … *(charted name)* on port side / starboard side.

.6 Buoy … *(charted name)* … is cables / nautical miles ahead.

.7 We are entering / we entered area ….

.8 We are leaving / we left area ….

.2 Movements

.1 True course / gyro compass course / magnetic compass course is … degrees.

.2 Gyro compass error is … degrees plus / minus.

.2.1 Magnetic compass error is … degrees east / west.

.3 Speed over ground / through water is … knots.

.4 Set and drift is … degrees, … knots.

.5 We are making … degrees leeway.

.6 The course board is written up.

.7 The next chart is within … hours.

.3 Draft

.1 Draft forward / aft is … metres.

.2 Present maximum draft is … metres.

.3 Underkeel clearance is … metres.

B1/1.2 Briefing on traffic situation in the area

.1 A vessel is

~ overtaking … *(cardinal points/half cardinal points)* of us.

~ on opposite course.

~ passing on port side / starboard side.

.2 A vessel is crossing from port side.

.2.1 The vessel

~ will give way.

~ has given way.

~ has not given way yet.

~ is standing on.

~ need not give way.

.3 A vessel is crossing from starboard side.

.3.1 We

~ need not give way.

~ will stand on.

~ will alter course to give way.

~ have altered course to give way.

.3.2 The vessel will pass … kilometres / nautical miles ahead / astern.

.3.3 I will complete the manoeuvre.

.4 A vessel … *(cardinal points/half cardinal points)* of us is on the same course.

.5 The bearing to the vessel in … degrees is constant.

.6 There is heavy traffic / … in the area.

.6.1 There are fishing boats / … in the area.

.7 There are no dangerous targets on the radar.

.7.1 Attention. There are dangerous targets on the radar.

.8 Call the Master if any vessel passes with a CPA of less than … miles.

.8.1 Call the Master if ….

B1/1.3 Briefing on navigational aids and equipment status

.1 Port side / starboard side radar is at … miles range scale.

.2 The radar is

~ relative head-up / north-up / course-up.

~ true-motion north-up / course-up.

.3 GPS / LORAN is / is not in operation.

.4 Echo sounder is at … metres range scale.

.4.1 The echo sounder recordings are unreliable.

.5 I changed to manual / automatic steering (at … hours UTC).

.6 Navigation lights are switched on / off.

B1/1.4 Briefing on radiocommunications

.1 INMARSAT … *(type of system)* is operational / is not operational.

.2 VHF DSC Channel 70 / VHF Channel … / DSC controller is switched on.

.2.1 DSC frequency 2187.5 kHz is switched on.

.3 NAVTEX is switched on.

.4 Following was received on … at … hours UTC.

.5 Shore based radar assistance / VTS / Pilot station is on VHF Channel ….

.6 The Pilot station / VTS station requires

~ flag State.

~ call sign / identification.

~ draft.

~ gross tonnage.

~ length overall.

~ kind of cargo.

~ ETA at ….

~ ….

B1/1.5 Briefing on meteorological conditions

.1 A weak / strong (tidal) current is setting … degrees.

.1.1 The direction of the (tidal) current will change in … hours.

.2 Fog / mist / dust / rain / snow / … is in the area.

.3 Automatic fog signal is switched on.

.4 The wind increased / decreased (within last … hours).

.4.1 The wind is … *(cardinal points/half cardinal points)* force Beaufort ….

.4.2 The wind changed from … *(cardinal points/half cardinal points)* to … *(cardinal points/half cardinal points)*.

.5 The sea state is expected to change (within … hours).

.6 A smooth/moderate/rough/high sea - slight/moderate/heavy swell of … metres from … *(cardinal points/half cardinal points)* is expected (within … hours).

.7 A tsunami / an abnormal wave is expected by … hours UTC.

.8 Visibility is … nautical miles.

.9 Visibility is reduced by fog / mist / dust / rain / snow / ….

.10 Visibility is expected

~ to decrease / increase to … nautical miles (within … hours).

~ variable between … and … nautical miles (within … hours).

.11 Next weather report is at … hours UTC.

.12 Atmospheric pressure is … millibars/hectopascals.

.13 Barometric change is … millibars/hectopascals per hour / within the last … hours.

.13.1 Barometer is steady / dropping (rapidly) / rising (rapidly).

.14 There was a gale warning / tropical storm warning for the area … at … hours UTC.

2

B1/1.6 Briefing on standing orders and bridge organization

.1 Standing orders for the period from … to … hours UTC … are: ….

.2 Standing orders for the area … are: ….

.3 Take notice of changes in the standing orders.

.4 Do you understand the standing orders?

.4.1 Yes, I understand the standing orders.

.4.2 No, I do not understand, please explain.

.5 Read / sign the standing orders.

.6 The latest fire patrol was at … hours UTC.

.7 The latest security patrol was at … hours UTC.

.7.1 Everything is in order.

.7.2 The following was stated: ….

.7.3 The following measures were taken: ….

.7.4 The following requires attention: ….

.8 The lookout is standing by.

.9 The helmsman is standing by.

.10 Call the Master at … hours UTC / in position ….

B1/1.7 Briefing on special navigational events

See also A1/3 "Safety communications".

.1 There was an engine alarm at … hours UTC due to ….

.2 Speed was reduced at … hours UTC due to ….

.3 Engine(s) was / were stopped at … hours UTC due to ….

.4 Course was altered at … hours UTC due to ….

.5 The Master / Chief Engineer was called at … hours UTC due to ….

B1/1.8 Briefing on temperatures, pressures and soundings

.1 The …(equipment) temperature minimum/maximum is

~ … degrees (centigrade) /to maintain.

~ … degrees above / below normal.

~ critical.

.1.1 Do not exceed a minimum/maximum temperature of … degrees.

.2 The …(equipment) pressure minimum/maximum is

~ … bars/to maintain.

~ above / below normal.

~ critical.

.2.1 Do not exceed a pressure of … kiloponds / bars.

.3 Ballast / fresh water / fuel / oil / slop sounding is … metres / cubic metres.

.3.1 Sounding of

~ no … cargo tank is … metres / cubic metres.

~ no … cargo hold is … centimetres.

~ ….

B1/1.9 Briefing on operation of main engine and auxiliary equipment

See also B1/1.8.

.1 (present) revolutions of the main engine(s) are … per minute.

.2 (present) output of the main engine(s) / auxiliary engine(s) are … kilowatts.

.3 (present) pitch of the propeller(s) is … degrees.

.4 There are no problems.

.5 There are problems ….

~ with the main engine(s) / auxiliary engine(s).

~ with ….

6 Call the watch engineer (if the problems continue).

.6.1 Call the watch engineer … minutes before the arrival at … / at … hours UTC.

B1/1.10 Briefing on pumping of fuel, ballast water, etc.

.1 There is no pumping at present.

.2 We are filling / we filled (no.) … double bottom tank(s) / the ballast tanks / the … tank(s).

.2.1 Fill up … tonnes / sounding … / ullage … / level … to the alarm point.

.3 We are discharging / we discharged (no.) … double bottom tank(s) / the ballast tanks / the … tank(s).

.4 We are transferring / we transferred fuel / ballast / fresh water / oil from (no.) … tank(s) to (no.) … tank(s).

.5 We require a further generator to operate an additional pump.

B1/1.11 Briefing on special machinery events

.1 There was a breakdown of the main engine(s) (at … hours UTC / from … to … hours UTC).

.1.1 There was a breakdown of … (at … hours UTC / from … to … hours UTC).

.2 There was a blackout (at … hours UTC / from … to … hours UTC).

.2.1 There was a blackout in … (at … hours UTC / from … to … hours UTC).

.3 Main engine(s) was / were stopped (at … hours UTC / from … to … hours UTC) due to ….

.4 Speed was reduced (at … hours UTC / from … to … hours UTC) due to ….

.5 Call the Master / Chief Engineer if the revolutions of the main engine(s) are below … per minute.

.5.1 Call the Master / Chief Engineer / Watch Engineer if ….

B1/1.12 Briefing on record keeping

.1 The log books / record books are completed and signed.

.1.1 The note book entries will be copied (into the log books / record books) after the watch.

.2 Change the paper of the data logger / echo sounder / … recorder.

.2.1 Refill the toner / ink of the data logger / echo sounder / … recorder.

B1/1.13 Handing and taking over the watch

The Master / Chief Engineer or an (engineer) officer handing over the watch should say:

.1 You have the watch now.

.1.1 The relieving officer should confirm and say:

I have the watch now.

The Master / Chief Engineer when called to the bridge / engine (control) room and formally taking over the watch, should confirm and say:

.2 I have the watch now.

.2.1 The officer of the watch should confirm and say:

You have the watch now.

Watch Keeping Duty Schedule

① 00～04時（12～16時）「ゼロヨン」 Second mate/officer

② 04～08時（16～20時）「ヨンパー」 Chief mate/officer

③ 08～12時（20～24時）「パーゼロ」 Third mate/officer

Lesson 3

Listening and Speaking

Learning Objectives

At the end of the lessons activities, the students are able to:

a) practice in the role play maritime conversation using the IMO SMCP and techical knowledge on specifical situation.

b) apply listening skill techinques and identify the correct answers in the listening test.

Everybody should speak in English. Navigation students will be the navigating officers of the Yuge Maru and the Engineering students will be the engineers.

Task : Assign the roles and the tasks to your group members and plan what to do in your role play.

At the Bridge: Captain, Third officer, AB or Helmsman

At the Forward Station: Chief officer, Boatswain, Line man or cadet

At the Aft Station: Second officer, Line man or cadet

At the Engine Control Room: First engineer or Second engineer

At the Engine Room: Oiler, Wiper or cadet

Goal : Students will be able to speak in English and use some maritime terms correctly while doing the task as ship's officers and crews.

○早口ことば(3.1)

学習目標

同じような発音が続く複雑な文章を正確かつスムーズに発音できるように学習しましょう。

学習方法

「2隻の船舶が交錯するときに守るべきルール」の早口ことばをペアをつくってスムーズに言えるようになるまで練習しましょう。

○ロールプレイ(3.2，3.3，3.4，3.5)

学習目標

機関制御室における当直交代(3.2)，消火活動(3.3)，錨巻き揚げ機のトラブルシューティング(3.4)，出航時の船橋と機関制御室の会話(3.5)の各シーンを想定して，ロールプレイを行い，実務に近い英語による会話ができるように学習しましょう。

学習方法

①まずグループに分かれ，船長，一等航海士，甲板員，機関長，一等機関士，操機手など，ロールプレイ上の一人一人の役割を決めましょう。

②各手順チェックリストを基に，実際に交わす会話をグループ全員で一緒に考え，練習しましょう。

- ほとんどの会話はSMCPによって国際標準化(決まり文句)されています。まずはSMCPを覚えましょう。
- 会話はゆっくり，はっきり発音することを心掛けましょう。
- オーダーを受けた役者は必ずアンサーバックする練習をしましょう。

③実際に練習船の船橋，甲板，機関制御室，機関室などを使いロールプレイを行いましょう。

- お互いの顔が見えないコミュニケーションは難しいものです。顔の表情がコミュニケーションのかなりの部分を占めているからです。また，無線機や電話では音声の完璧な再生は不可能です。必ず歪や雑音が伴います。そこで，実際と同様に無線機や電話を用いて各部署とのコミュニケーションを行ってみましょう。
- 後で復習できるようにビデオ撮影をしておきましょう。

④撮影したビデオを見ながら，良かった点や改善すべき点について，他のグループのメンバーを交えて話し合い，次のロールプレイにフィードバックし，より良いものにしていきましょう。

○いろいろな出身国の人が話す英語（3.6）

学習目標

一口に英語といっても実際にはいろいろな英語があります。ネイティブでも，イギリス人が話す英語とアメリカ人が話す英語は違います。ちょうど日本語でも関東の人が話す日本語と関西の人が話す日本語が違うような感じです。ましてや，英語を母国語としない国の人たちが話す英語はかなり違います。外航船には日本人以外に，フィリピン，インド，インドネシア，クロアチアの人たちが相当数乗っています。船長・機関長以外はすべて外国人という状況も多々あります。これらの乗組員と船舶の運航に必要なコミュニケーションを円滑にとらなくてはなりません。そこで，いろいろな出身国の人が話す英語の癖を理解し，円滑なコミュニケーションが行えるように，ここで学習しましょう。

学習方法

①イギリス人，インド人，日本人が話す会話を聞いて設問に答えてください。
（会話のwavファイルはhttp://www.kaibundo.jp/me.htmからダウンロードできます）
②それぞれの会話がどこの国の人によって話されているか答えてください。

○トラブルシューティング（3.7）

学習目標

船でトラブルが発生したときに，適切な状況確認を行い，適切な対処ができるように練習しましょう。

学習方法

問題文に書かれているトラブルが発生したときに確認すべき点を，グループで話し合いましょう。

3.1 Twist Your Tongue and Thinking Exercise

Ⓝ

The first mate of the ferry found the full speed ahead in the fairway too furious as their ferry is too formidably old.

If the first mate found the ferry too formidably old, why did the first mate find the ferry to its full speed ahead?
So, when the fast ferry furiously found her way fast, why did the first mate worry for the ferry to make fast to the shore?

3.2 Hand Over of the Watch E

Role Play

The following things need to be informed to the reliving officer:

① **Special orders** related to any ship operation, control system, or maintenance work.
② **Standing orders** from the chief engineer or the company.
③ Level of **important tanks** such as bilges, ballast tank, sewage tank, reserve tank, slop tank, fuel tank, or any other tank which requires attention.
④ Condition and state of **fire extinguishing equipment** and systems, in case any specific section or fire alarm has been isolated.
⑤ **Special mode of operation** in case of emergency situation, damage, icy, or shallow water etc.
⑥ In case there is any kind of **maintenance work** being carried out in the engine room by other engineer officer and crew members, then their work location, details of machinery under maintenance, and information of authorized person and crew members should be provided. Any potential hazard because of the ongoing maintenance work should also be informed.
⑦ In case there is an **equipment failure**, details of the same should be informed.
⑧ **All the checks** already made when the ship leaves the port should be noted. In case any check is pending, it should be conveyed to the reliving officer.
⑨ All the checks that are made when the ship enters the port should be noted and informed in case any is missing.
⑩ Condition and **important information regarding mode of operation** of main engine, boiler, and auxiliary engines should be informed.
⑪ In case an **equipment needs to be monitored manually**, details of the same should be provided, along with the condition of monitoring and control equipment.
⑫ Any form of **adverse ship condition** needs to be informed.
⑬ Information on the condition and modes of all the **important auxiliary machinery** such as purifiers, fresh water generator, oily water separator, pumps, sewage treatment plant, etc. should be provided.
⑭ In case any **important machinery failed** to receive attention during the watch, the reliving officer should be reported and asked to take care of the same.
⑮ The condition and **modes of automatic boiler** controls and details of other equipment related to the operation of the steam boiler should be provided.
⑯ The engineer officer should ensure that **all the important parameters** regarding main and auxiliary machines are suitably recorded in the engine room log book.

Practice

List down all the important words that are used in every briefing to be done by the outgoing duty engineer to the incoming watch duty engineer.

3.3 Fire Fighting

N E

Role Play
Work in Pairs

Take a look at the SMCP dialogues on fire fighting and drills. From the four sub-sections (reporting fire, reporting readiness for action, orders for fire fighting and cancellation of alarm), write a script on fire including the four scenes. Indicate the speaker of each speaking line.

e.g. Reporting Fire
AB：Fire onboard!
Smoke in hold number 2.
CM：Report injured person.
AB：No person injured.

B2/3.2 **Fire fighting and drills**

.1 Reporting fire

.1 Fire on board!

.1.1 Smoke / fumes / fire / explosion
～ in engine room.
～ in no. … hold(s) / tank(s).
～ in superstructure / accommodation.
～ in … space.
～ on deck / ….

.1.2 Smoke / fumes from ventilator(s).

.1.3 Burnt smell / fumes in … / from ….

.2 Report injured persons / casualties:

.2.1 No person injured.

.2.2 Number of injured persons / casualties is: ….

.3 What is on fire?

.3.1 Fuel / cargo / car(s) / truck(s) / wagon(s) / containers (with dangerous goods) / … on fire.

.3.2 No information (yet).

.4 Is smoke toxic?

.4.1 No, smoke not toxic.

.4.2 Yes, smoke toxic

.5 Is fire under control?

.5.1 Yes, fire (in …) under control.

.5.2 No, fire (in …) not under control (yet).

.5.2.1 Fire spreading (to …).

.5.2.2 Fire (in …) not accessible.

.6 Report damage.

.6.1 No damage.

.6.2 Minor / major damage in … / to ….

.6.3 No power supply (in …).

.6.4 Making water in ….

.7 Pressure on fire mains!

.8 Shut down main engine(s) / auxiliary engine(s) / … and report.

.8.1 Main engine(s) / auxiliary engine(s) / … shut down.

.9 Stop fuel and report.

.9.1 Fuel stopped.

.10 Close all openings (in … / in all rooms) and report.

.10.1 All openings (in … / in all rooms) closed.

.10.1.1 Openings in … not accessible.

.11 Switch off ventilator(s) (in …) and report.

.11.1 Ventilator(s) (in …) switched off.

.12 Turn bow / stern to windward.

.13 Turn port side / starboard side to windward.

.14 Alter course to ….

.2 Reporting readiness for action

.1 Stand by fire fighting team / rescue team / first aid team / support team and report.

.1.1 Fire fighting team / rescue team / first aid team / support team standing by.

.2 Stand by main engine and report.

.2.1 Main engine standing by.

.3 Stand by CO_2 station / … station / emergency generator.

.3.1 CO_2 station / … station / emergency generator standing by.

.4 Close all openings (in … / in all rooms) and report.

.4.1 All openings (in … / in all rooms) closed.

.4.1.1 Openings in … not accessible.

.3 Orders for fire fighting

.1 Start fire fighting.

.1.1 Take one / two / … fire fighting teams / … team(s) to scene.

.2 Go following route:

.2.1 Go through engine room / no. … hold(s) / tank(s) / superstructure / accommodation / … space / manhole(s) to … space / funnel / ….

.2.2 Go from

~ outside / inside to ….

~ port side / starboard side to ….

~ … to ….

.3 Take following (additional) safety measures and report.

.3.1 Have two / … members in one team.

.3.1.1 Number of members in fire fighting team / … team is: ….

.3.2 Have lifeline between each other / to outside.

.3.2.1 … team members have lifelines to each other.

.3.2.2 … team has lifelines to outside.

.3.3 Have rescue team on stand-by and report.

.3.4 Maintain visual contact / radio contact on walkie-talkie.

.4 Fire fighting team must have following outfit:

.4.1 Fire fighting team must have protective clothing / smoke helmets / breathing apparatus / ….

.5 Manning of fire fighting team / … team(s) as follows:

.5.1 Chief Officer / Chief Engineer / … in command of fire fighting team / … team (no. …).

.5.2 Following officer(s) / crew member(s) in fire fighting team / … team: ….

.6 Restrict action (in … / on …) to … minutes.

.6.1 Agree on retreat signal and report.

.6.1.1 Retreat signal for fire fighting team / … team … is ….

.7 Use water / foam / powder / CO_2 / sand / … in ….

.8 Run out fire hoses and report.

.8.1 Fire hoses run out.

.9 Water on!

.9.1 Water is on.

.10 Cool down … with water and report.

.10.1 … cooled down.

.4 Cancellation of alarm

.1 Is the fire extinguished?

.1.1 Yes, fire (in …) extinguished.

.1.2 No, fire (in …) not extinguished (yet).

.1.3 Fire restricted to … space / area.

.2 Post a fire watch and report.

.2.1 Fire watch posted (in … space / area).

.3 Fire extinguishing systems / means remain on stand-by.

.4 Fire fighting team / … team remain on stand-by.

.5 Rope off the fire area and report.

.5.1 Fire area roped off.

.6 Check the fire area every … minutes / hour(s) for re-ignition and report.

.6.1 Fire area checked, no re-ignition.

.6.2 Fire area checked, re-ignition in … space / area.

.6.2.1 Re-ignition extinguished.

.7 The fire alarm is cancelled (with following restrictions: …).

3.4 Windlass Trouble upon Leaving the Anchorage Area

N E

3

Role Play

Chemical Tanker M/T Tina Theresa (mock up with T/S OSHIMA MARU CALL SIGN XXXX) has dropped anchor at outer anchorage area in Singapore. After 22 hours of waiting for berth at the Petronas Terminal, the Singapore harbor called up the chemical tanker ship and advised that M/T Tina Theresa can now enter the harbor at 1000H and proceed to berth 23. It was then 0830H.

Master/Captain of the tanker ship has advised the ER to standby engine 1 hour before heaving up anchor at 0930H. One hour later. Deck hands (C/M, Bosun and AB) are standing by at the forward part of the ship ready to heave up anchor.

Master (from the bridge using portable radio) ordered the C/M to heave up anchor and secure it. C/M ordered bosun to heave up, wash up the anchor and secure. As the Bosun and AB removed chain stopper and disengaged brake. There was slow move. Windlass conked out and no more response. C/M report to Master that windlass has trouble that they could not heave up anchor. Master advises the ER to send engineer to check the problem as quick as possible as the terminal is calling their ship already to proceed to berth.

NOTE : Engineers try to troubleshoot (Suggest correct action to the problem)

Maybe this info can help but this is non-conclusive. Assess the problem and solve it.

No response typically means a complete loss of power IE fuse, a solenoid had failed, or the small control circuit wires have come adrift. Check for control voltage at the solenoid control lugs. A good reading of control voltage means the wires and supply circuit are working, but the solenoid is not passing the voltage or hydraulic pressure.

3.5 Bridge and Engine Simulation on Pre-Departure Procedures N E

Role Play

YUGE MARU prepares to sail to Matsuyama. The captain gives order to the Third Officer to inform the 1 hour notice of departure to the Engine Department. (Call engine department)

Engine department acknowledges the call and prepares the engine. Deck department also prepares the ship's departure. (Do pre-departure procedures)

After the ship's preparation, Yuge Maru will sail. (Officers take the ship to navigation)

NOTE :

- All must be able to communicate properly while doing the tasks using the communication equipment onboard Yuge Maru.
- All the procedures in the Bridge and Engine must be performed correctly. (See Appendices)

YUGEMARU Departure Procedure (E/R)

From ____________________ To ____________________ Date ____________________

Name ____________________________

Warm-up Procedure

Procedure	Report
☐ Check C.F.W. quantity (Expansion TK)	· Water quantity of the Exp' TK is ________L.
☐ Check M/E Crankcase L.O. quantity	· We'll check the L.O. quantity of the M/E Crankcase.
☐ M/E L.O. S/B PP, "Start"	· We'll start the M/E L.O. S/B PP.
☐ M/E L.O. Heater, "Start"	· We'll start the M/E L.O. Heater.
☐ M/E C.F.W. S/B PP, "Start"	· We'll start the M/E C.F.W. S/B PP.
☐ M/E C.F.W. Heater, "Start"	· We'll start the M/E C.F.W. Heater.

YUGEMARU Departure Procedure

☐ Check C.F.W. quantity (Expansion TK)	· Water quantity of the Exp' TK is ________L.
☐ Check M/E Crankcase L.O. quantity	· We'll check the L.O. quantity of the M/E Crankcase.
Open Each Valves	
☐ F.O. SERVICE TK Inlet valve, "Open"	· We'll open the Inlet valve of the F.O. SERVICE TK.
☐ D/G F.O. Flow meter Inlet valve, "Open"	· We'll open the Inlet valve of the D/G F.O. Flow meter.
☐ M/E F.O. Flow meter Inlet valve, "Open"	· We'll open the Inlet valve of the M/E F.O. Flow meter.
☐ S.W. Port or St'b suction valve, "Open"	· We'll open the S.W. Port or St'b suction valve.
☐ M/E Overboard discharge valve, "Open"	· We'll open the Overboard discharge valve of the M/E.
☐ NO.1 and NO.2 D/G Overboard discharge valves, "Open"	· We'll open the NO.1 and NO.2 D/G Overboard discharge valves.
☐ Stern tube S.W. Inlet valve, "Open"	· We'll open the Inlet valve of Stern tube S.W.

Turning

☐ Engage the Turning Gear	· We'll engage the Turning Gear.
☐ Turning Motor power, "ON"	· We'll turn on the Turning Motor.
☐ Turning, "Start" (Check the current value)	· We'll start the turning. The current is ______A.

3

Start Main Air Compressor

☐ Main Air Reservoir drain discharge	· We'll discharge blow off the drain of the NO.1 or NO.2 Main Air Reservoir.
☐ Main stop valve, "Open"	· We'll open the Main stop valve.
☐ Main Control Air valve, "Open"	· We'll open the Main Control Air valve.
☐ Charge valve, "Open"	· We'll open the Charge valve.
☐ Main Air Compressor breaker, "OFF"	· We'll start the Main Air Compressor.

Start NO.1, NO.2 D/G

☐ Check L.O. SUMP TK quantity	· We'll check the L.O. quantity of the D/G SUMP TK.
☐ Check Starting Air Motor L.O. quantity	· We'll check the Starting Air Motor.
☐ Check C.F.W. TK quantity	· We'll check the C.F.W. TK quantity.
☐ Priming	· We'll carry out L.O. Priming for the D/G.
☐ D/G, "Start"	· We'll start the D/G.
☐ CPP Oil PP, "Start"	· We'll start the CPP Oil PP.
☐ R/G L.O. PP, "Start"	· We'll start the R/G L.O. PP.
☐ M/E S/B F.O. SUPPLY PP, "Start"	· We'll start the M/E S/B F.O. SUPPLY PP.

End Warm-up

☐ M/E L.O. Heater, "Stop"	· We'll stop the M/E L.O. Heater.
☐ M/E C.F.W. Heater, "Stop"	· We'll stop the M/E C.F.W. Heater.
☐ M/E Turning, "Stop"	· We'll stop turning.
☐ Turning Motor power, "OFF"	· We'll turn off the Turning Motor.
☐ Disengage the Turning Gear	· We'll disengage the Turning Gear.

Air Running

☐ Starting Air Inlet valve, "Open"	· We'll open the Starting Air Inlet valve.
☐ Request to C/R (Starting Air Intermediate valve, "Open")	· Please open the Starting Air Intermediate valve.
☐ Drain valve, "Close"	· We'll close the Drain valve.
☐ Check the F.O. handle STOP position	· We'll check the F.O. handle STOP position.
☐ Air running	· We'll carry out air running for the M/E.
☐ Check the abnormality	· Everything is in order, Sir.

Start M/E

☐ Indicator valve, "CLOSE"	· Plase close all indicator valves on the M/E.
☐ F.O. Handle, "RUN"	· We'll change position of F.O. control handle from STOP to RUN.
☐ Report to C/R	· Engine room all's well.
☐ Started the M/E (550rpm), Check the abnormality	· Everything is in order, Sir.
☐ M/E S/B F.O. SUPPLY PP, "Stop"	· We'll stop the M/E S/B F.O. SUPPLY PP.
☐ M/E L.O. S/B PP, "Stop"	· We'll stop the M/E L.O. S/B PP.
☐ M/E C.F.W. S/B PP, "Stop"	· We'll stop the M/E C.F.W. S/B PP.
☐ R/G L.O. PP, "Stop"	· We'll stop the R/G L.O. PP.

Engage Clutch

☐ C.S.W. PP "Start" (Cheak the discharge pressure)	· We'll start the C.S.W. PP.
☐ After increase of M/E revolution (～750rpm), Check for abnormality	· Everything is in order, Sir.

YUGEMARU Departure Procedure (C/R)

From ____________________ To ____________________ Date ____________________

Name ____________________________

Procedure	Report
☐ Lamp test	· We'll check the Lamp of the M/E console.
Change Inboard Power	
☐ Check D/G Start、Exh' Gas Temp', Voltage and Frequency	· We'll check both D/G Start. the Exh' Gas Temp' is ______degrees Celsius. the Voltage is ______V, the Frequency is ______Hz.
☐ NO.1 or NO.2 D/G ACB, "CLOSE"	· We'll close NO.1 or NO.2 D/G ACB.
☐ CPP Control box Power, "ON"	· We'll turn on the CPP Control box.
☐ SEC Power, "ON"	· We'll turn on the SEC.
D/G Parallel Operation	
☐ Switch Synchro & load sharing AUTO→MANU	· We'll switch the Synchro & load sharing from AUTO to MANU.
☐ Adjust synchroscope→NO.1 or NO.2 D/G	· We'll adjust the synchroscope to NO.1 or NO.2 D/G.
☐ Adjust Phase & Frequency (Use governor)	· We'll adjust the Phase and Frequency using a governor.
☐ Reading Phase, ACB, "CLOSE"	· We'll close ACB.
☐ Synchroscope, "OFF"	· We'll turn off the Synchroscope.
☐ Adjust Load sharing (Use governor)	· We'll adjust the Load sharing using a governor.
☐ Switch Synchro & load sharing MANU→AUTO	· We'll switch Synchro & load sharing from MANU to AUTO.
☐ Thruster power, "ON"	· We'll turn on the Thruster power.
☐ Report to W/H	· We turned on the Thruster power.
Air Running	
☐ confirm with E/R→Starting Air Intermediate valve, "Open"	· We'll open the Starting Air Intermediate valve.
☐ CPP move test	· The change over of CPP Blade angle has been tested.
M/E Start	
☐ M/E Start ＊Start time (______:______)	· We'll start the M/E.
Clutch Engage	
☐ Check start & pressure of S.W. PP	· We started the S.W. PP. Pressure is ______Pa.
☐ Engage Clutch ＊Time (______sec)	· We'll engage the Clutch.
☐ Increase RPM (550→750)	· We'll increase the RPM from 550 to 750.
☐ On report from W/H→Set Thruster power, "OFF"	· We'll turn off the Thruster.
D/G Single Operation	
☐ Switch Synchro & load sharing AUTO→MANU	· We'll switch the Synchro & load sharing from AUTO to MANU.
☐ Adjust After load (Use governor)	· We'll adjust the After load using a governor.
☐ ACB of D/G to be stopped, "Open"	· We'll open the ACB of NO.1 or NO.2 D/G.
☐ D/G, "Stop"	· We'll stop the NO.1 or NO.2 D/G.
☐ Check READY TO START light	· READY TO START light is on.
☐ Check S/B SELECT→NO.1 or NO.2 D/G	· We'll check the S/B SELECT to NO.1 or NO.2 D/G.
☐ Switch Synchro & load sharing AUTO→MANU	· We'll switch Synchro & load sharing from MANU to AUTO.

YUGEMARU Entering Port Procedure (E/R)

From ____________________ To ____________________ Date ____________________

Name ____________________________

YUGEMARU Entering Port Procedure

Procedure	Report
Start Each PP	
☐ M/E C.F.W. S/B PP, "Start"	· We'll start the M/E C.F.W. S/B PP.
☐ M/E L.O. S/B PP, "Start"	· We'll start the M/E L.O. S/B PP.
☐ R/G L.O. S/B PP, "Start"	· We'll start the R/G L.O. S/B PP.

After Engine Finish

Procedure	Report
Air Blow	
☐ Check F.O. handle, STOP position	· We'll check the F.O. handle is at STOP position.
☐ Indicator valve, "Open"	· Open the Indicator valve of the M/E.
☐ Air Blow	· We'll carry out Air Blow of the M/E.
☐ Check everything is in order (Foreign matter, Water etc…)	· Everything is in order, Sir.
☐ Starting Air Inlet valve, "Close"	· We'll close the Starting Air Inlet valve.
☐ Drain valve residual pressure blow (Drain valve kept open)	· We'll blow off the Drain valve.
Turning	
☐ Engage Turning Gear	· We'll engage the Turning Gear.
☐ Turning Motor power, "ON"	· We'll turn on the power for the Turning Motor.
☐ Turning, "Start" (Check current)	· We'll start turning. The current value is ______A.
☐ CPP Oil PP, "Stop"	· We'll stop CPP Oil PP.
☐ Check the M/E J.C.F.W. outlet temp', 60℃ or less	· The J.C.F.W. outlet temp' of the M/E is less then 60 degrees Celsius.
Stop Each PP	
☐ M/E L.O. S/B PP, "Stop"	· We'll stop the M/E L.O. S/B PP.
☐ M/E C.F.W. S/B PP, "Stop"	· We'll stop the M/E C.F.W. S/B PP.
☐ R/G L.O. S/B PP, "Stop"	· We'll stop the R/G L.O. S/B PP.
Close Each S.W., F.O., Air Valve	
☐ Stern tube S.W. Inlet valve, "Close"	· We'll close the Inlet valve of the Stern tube S.W.
☐ F.O. SERVICE TK Inlet valve, "Close"	· We'll close the Inlet valve of F.O. SERVICE TK.
☐ D/G F.O. Flow meter Inlet valve, "Close"	· We'll close the Inlet valve of the D/G F.O. Flow meter.
☐ M/E F.O. Flow meter Inlet valve, "Close"	· We'll close the Inlet valve of the M/E F.O. Flow meter.
☐ S.W. Port or St'b sea chest, "Close"	· We'll close valve of S.W. Port or St'b sea chest.
☐ M/E Overboard discharge valve, "Close"	· We'll close the Overboard discharge valve of the M/E.
☐ NO.1 and NO.2 D/G Overboard discharge valves, "Close"	· We'll close the NO.1 and NO.2 D/G Overboard discharge valves.
☐ Main stop valve, "Close"	· We'll close the Main stop valve.
☐ Control Air valve, "Close"	· We'll close the Control Air valve.
☐ After air being charged→Charge valve, "Close"	· We'll check the Air Charge after closing the Charge valve.
☐ Main Air Compressor breaker, "OFF"	· We'll turn off the Main Air Compressor.

YUGEMARU Entering Port Procedure (C/R)

From ____________________ To ____________________ Date ____________________

Name ____________________

YUGEMARU Entering Port Procedure

Procedure	Report
D/G Parallel Operation	
☐ Switch Synchro & load sharing AUTO→MANU	· We'll switch the Synchro & load sharing from AUTO to MANU.
☐ Adjust synchroscope→NO.1 or NO.2 D/G	· We'll adjust the synchroscope for NO.1 or NO.2 D/G.
☐ Adjust Phase and Frequency (Use governor)	· We'll adjust the Phase and Frequency using a governor.
☐ Looking Phase, ACB, "CLOSE"	· We'll close ACB.
☐ Synchroscope, "OFF"	· We'll turn off the Synchroscope.
☐ Adjust Load sharing (Use governor)	· We'll adjust the Load sharing using a governor.
☐ Switch Synchro & load sharing MANU→AUTO	· We'll switch the Synchro & load sharing from MANU to AUTO.
☐ Thruster power, "ON"	· We'll turn on the Thruster power.
☐ Report to W/H	· We turned on the Thruster power.

After Engine Finish

Procedure	Report
☐ Decrease RPM (750rpm→550rpm)	· We'll decrease the RPM from 750 to 550.
☐ Disengage clutch	· We'll disengage the clutch.
☐ Stop the propeller shaft (______sec)	· Stopping the propeller shaft took ______sec.
☐ Check M/E EXH' Gas Temp', 200℃ or less	· The EXH' Gas Temp' of the M/E became less than 200 degrees Celsius.
☐ Stop M/E ＊Stop time (______:______)	· We'll stop the M/E.
☐ Starting Air Intermediate valve, "Close"	· We'll close the Starting Air Intermediate valve.
☐ Thruster power, "OFF"	· We'll turn off the Thruster power.
☐ D/G Single Operation (Check READY TO START)	· We'll change over the D/G to Single Operation.
Power OFF	
☐ M/E remote power, "OFF"	· We'll turn off the M/E remote power.
☐ CPP Control unit power, "OFF"	· We'll turn off the CPP Control unit power.
☐ SEC power, "OFF"	· We'll turn off the SEC power.
Time using Shore Connection	
☐ Check Shore connection cable connect, voltage and frequency	· We'll check the Shore connection cable. Voltage is ______V, Frequency is ______Hz.
☐ Shore connection breaker, "ON"	· We'll turn on Land power.
☐ Shore connection supply (SHORE) ACB, "CLOSE" (D/G ACB is auto open)	· We'll close Shore connection supply (SHORE).
☐ D/G, "Stop"	· We'll stop the D/G.

3.6 Listening Test

N E

Listen carefully to the text you will hear. It consists of 24 multiple choice questions in English language which you will hear **twice**. Choose the correct answer to the question by circling the letter of your answer.

3

Set A

① (a) in drums
(b) in packages
(c) in bulk

② (a) crude oil in bulk
(b) water ballast
(c) the ship's engine and boilers

③ (a) oil-tight and water-tight
(b) connected to the loading arms
(c) fitted with a door

④ (a) electric motors
(b) valves and pumps
(c) levers and knobs

⑤ (a) amidships
(b) on the poop deck
(c) on the forecastle

⑥ (a) steeling
(b) framing
(c) hulling

⑦ (a) bulk heads
(b) side walls
(c) cofferdams

⑧ (a) the navigating bridge
(b) the engine room
(c) the cargo tanks

Set B

① (a) operator
(b) terminal
(c) engineer

② (a) pump out
(b) top up
(c) top off

③ (a) pumping pressure
(b) by-pass valve
(c) delivery suction

④ (a) called by the duty mate
(b) called by the terminal
(c) called by the AB

⑤ (a) radio
(b) checklist
(c) detector

⑥ (a) by-pass valve
(b) control valve
(c) control lever

⑦ (a) leaking
(b) pressure gauge
(c) pipeline

⑧ (a) heavy fine
(b) ship detained
(c) imprisonment

Set C

① (a) channelling
(b) arrival
(c) transiting

② (a) ship to shore
(b) shore to ship
(c) ship to ship

③ (a) MT Tristan
(b) MV Tristan
(c) MS Tristan

④ (a) Ro-Ro
(b) car ferry
(c) car carrier

⑤ (a) 37° 14' N; 122° 28.6' W
(b) 37° 14.4' N; 122° 28' W
(c) 37° 14.4' N; 122° 28.6' W

⑥ (a) 1100h
(b) 1030h
(c) 1000h

⑦ (a) 17 knots
(b) 16.3 knots
(c) 17.3 knots

⑧ (a) 3-4-2
(b) 3-2-4
(c) 3-4-4

(1) Guess the nationality of the three speakers who read the test:

Set A ______________________________

Set B ______________________________

Set C ______________________________

(2) Which speaker you found it most difficult to easiest to understand? (Rank by speaker A, B or C)

(most difficult) ___________ (average difficult) ___________ (least difficult) ___________

(3) Which test form you found most difficult to least difficult? (Rank 1 as easy and 3 difficult)

Set A _______ Set B _______ Set C _______

3.7 Basic Work / Troubleshooting E

Check for	Possible Cause/s	Remedy/ies
① Ballast pump failed		
② Mooring wintch failed		
③ Change of RPM		
④ Navigation lights fused		
⑤ Sewage plant failed		
⑥ Funnel sparks		
⑦ Hatch cover failed		
⑧ Hydraulic hose burst		
⑨ Low suction of pump		
⑩ Low discharge		
⑪ Overheat of motor		
⑫ Air bottle not filling air		
⑬ 110 volts grouded		
⑭ Bilge well high level		
⑮ Misfire of boiler		

Lesson 4

Reading and Writing

Learning Objectives

At the end of the lessons activities, the students are able to :

a) understand what is danger in illustration and what to do to prevent danger.
b) understand story and answer question.
c) write and read Navigation Log Book and Engineer's Log Book.
d) understand what is basic safety device in illustration and how to use it.

○イラスト(4.1，4.2，4.3)

学習目標

イラストを見て，どんな危険が潜んでいるか，どのような注意をすれば事故が防げるかを作文できることを目標に学習します。

学習方法

①イラストを見て，どんな危険が潜んでいるか作文しましょう。

②どのような注意をすれば事故が防げるか作文しましょう。

○4コマ漫画(4.4)

学習目標

4コマ漫画を読んで状況に合った内容の説明を書けるようになりましょう。

学習方法

①話の内容に関する質問に答えて，話の内容が理解できたか確認しましょう。

②話のなかに出てくる海事英単語を学習しましょう。

③役割を決めて，会話を考えましょう。次に実際にロールプレイをしてみましょう。

○航海日誌(4.5)

学習目標

航海日誌は航海の概要を英語で記入することが義務づけられている大切な書類です。正確に記入できるよう学習します。

学習方法

①航海日誌の記入項目，風力階級，波浪階級，天気などの航海日誌の記入に必要な英語表現を学習しましょう。

②桟橋へ行き，練習船のドラフトを計測し，航海日誌に記入しましょう。

③実際の航海実習のときに，航海コースを航海日誌に記入しましょう。

○機関日誌（4.6）

学習目標

機関日誌は機関の運転状況を英語で記入することが義務づけられている大切な書類です。正確に記入できるように学習します。

学習方法

①機関部署配置，機関停止，機関部署解散，出港，入港などの機関日誌の記入に必要な英語表現を学習しましょう。

②航海例の機関日誌から，航海時間，航進時間，機関使用時間，スリップを計算し，英語で機関日誌に記入できるように学習しましょう。

○安全設備（4.7）

学習目標

船舶に装備されている安全設備の名称と働きを理解して，非常事態のときに適切に安全設備を使用できるようにする。

学習方法

写真に示す安全設備がどのようなものか英語で説明しましょう。

4.1 The Points of Hazards

N E

① Label the things you know in the picture.
② What are dangers that might happen?
③ Explain your answer.

Picture 1

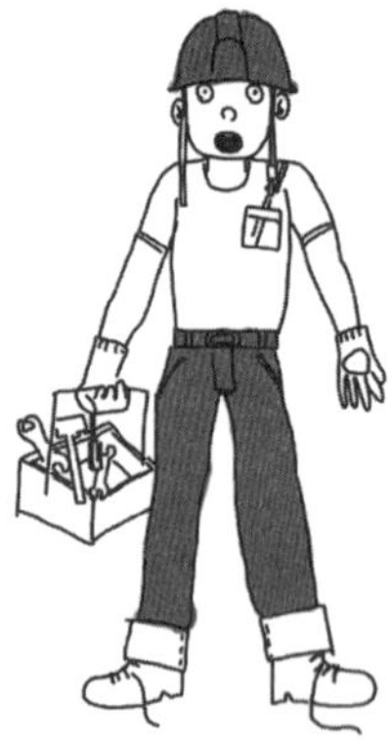

Picture 2

Picture 3

Picture 4

Picture 5

Picture 6

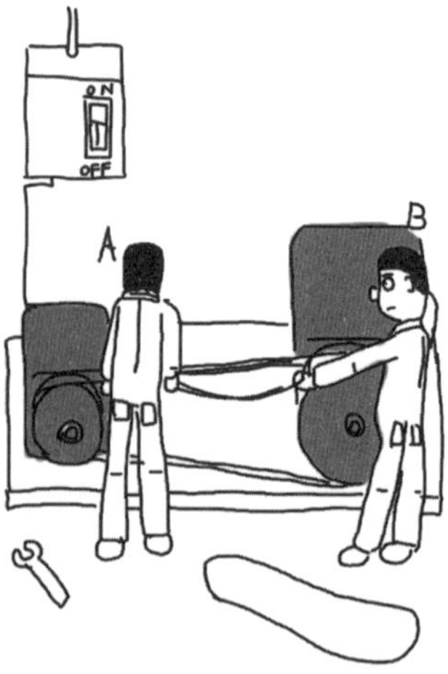

Picture 7

Picture 8

Picture 9

Picture 10

Picture 11

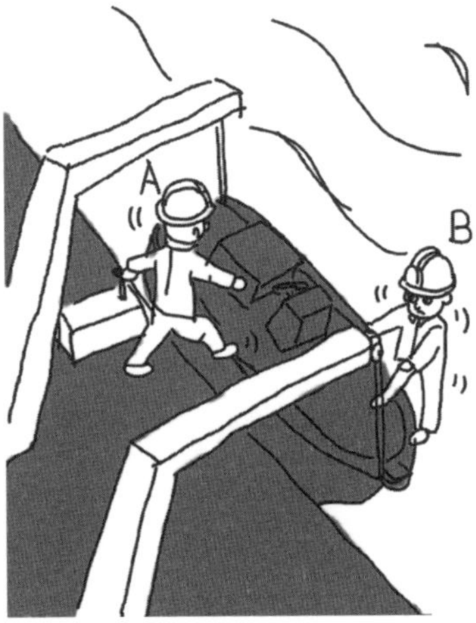

Picture 12

Picture 13

Picture 14

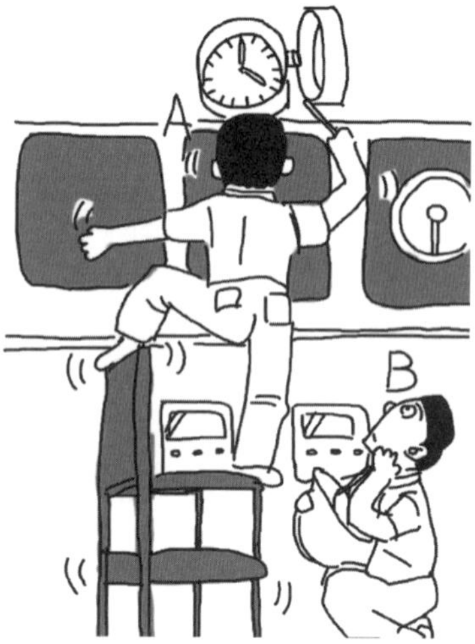

Picture 15

Picture 16

Picture 17

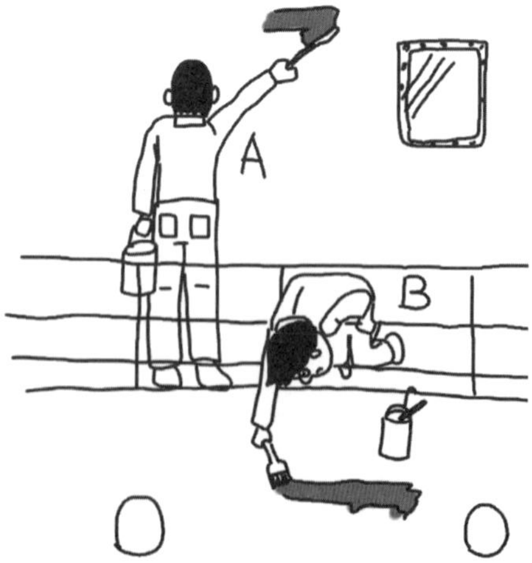

4.2 What is the Danger? N

Think about the situations below. Tell what is/are the wrong practice in every picture.

①

②

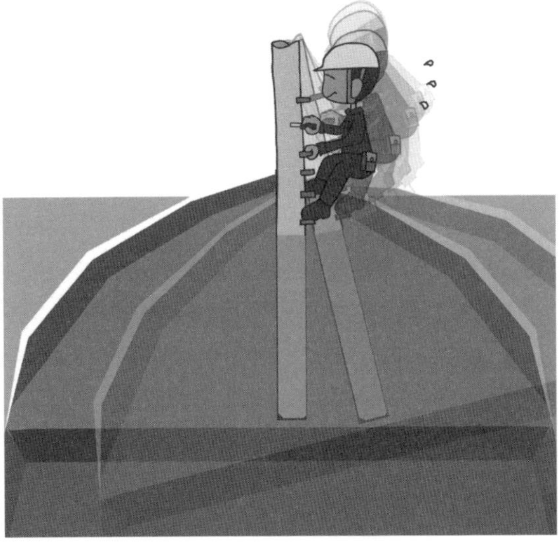

③

4.3 What is the Danger? E

Think about the situations below. Tell what is/are the wrong practice in every picture.

①

②

③

4.4 Situational Pictures for Conversation N E

Story 1 : Fire Fighting

1
Fire!
Fire on the main deck!
2
Master,
fire on the main deck!
Sound the alarm and
crews muster.
3
Deck department fight the fire.
Engine department cool
the bundry.
4
Ready forward stop!

1 Understanding the Story

① What is the situation in the story? ______________________

② What does the seaman do in the first picture? ______________________

③ Who are talking in the second picture? ______________________

④ Where did the crews gather in picture 3? ______________________

⑤ What did the crews do in picture 4? ______________________

2 Learning Vocabulary

What do you mean by these words? Look them up in the dictionary.

① Main deck ______________________

② Alarm ______________________

③ Crew ______________________

④ Muster station ______________________

⑤ Boundary cooling ______________________

3 Speaking Activity

Role play the fire situation. The students are assigned roles to do. Add more speaking lines in each situation.

Story 2 : The Seaman at the Airport

1 Understanding the Story

① What is the problem of the seaman? ______________________

② Where did he lose his ticket? ______________________

③ What place the seaman will go to? ______________________

④ What did the airline attendant do? ______________________

2 Learning Vocabulary

What do you mean of these words?

① A seaman signs-off from his ship. What does *signs-off* mean?

A. Put signature B. Disembark C. Write D. Complain

② *Attendant* means ______________________.

A. Crew B. Seller C. Bystander D. Cleaner

③ *Reprinted* is closest to ______________________.

A. Attended B. Made C. Looked for D. Contacted

3 Speaking Activity

Get a classmate who will be a partner in acting out the situation. Read the speaking lines and add more speaking lines.

Story 3 : The Seaman in the Department Store

1 Understanding the Story

① What did the seaman want to buy? ____________________

② What did he ask to the staff? ____________________

③ What did the manager do? ____________________

④ How much was the discount? ____________________

2 Learning Vocabulary

Look up the meaning of these words in the dictionary.

① Discount ____________________

② Request ____________________

③ Customer ____________________

3 Speaking Activity

Write your own script on buying a PSP. The store has many models so you talk to the salesman on the differences of each model for you to decide which you will buy.

Act out your script with a classmate in front of the class.

Story 4 : The Seaman Lost His Way

1 Understanding the Story

① What happened to the seaman?

A. Introduced himself to the attendant　　B. Lost his way to the next gate

C. Wanted to go the canteen

② Where is he going?

A. Taiwan　　B. Germany　　C. Amsterdam

③ What airline the seaman is flying?

A. JAL　　B. KLM　　C. ANA

2 Learning Vocabulary

Look up the meaning of these words in the dictionary.

① Attendant ______________________________

② Transit ______________________________

③ Signage ______________________________

3 Speaking Activity

Look from any old newspaper or magazine a comic illustration on any situation. Cut it out and paste it on a big paper. Make a story of what the comic illustration is. Present your work in front of the class.

Story 5 : The Seaman Complains on His Food

1 Understanding the Story

① Where is the seaman in the situation? ______________________

② What kind of food did the seaman order? ______________________

③ What kind of food was served to him? ______________________

④ Did he like the food? ______________________

2 Learning Vocabulary

Look up the meaning of these words in the dictionary.

① Recommend ______________________

② Cuisine ______________________

③ Exotic ______________________

3 Speaking Activity

Pretend that you are a waiter who will convince a customer to try your best-selling food in the restaurant. Role play the situation with a classmate.

4.5 Navigation Log Book Writing Ⓝ

Go to the taring ship and check items which you need to fill in the following navigation log book.

Hours	Miles	Tenths	Gyro Course	Gyro Error	Wind		Weather	Barometer	Temperature	
					Direction	Force			Air	Sea
1										
2										
3										
4										
11										
Noon										
23										
M.N										

	Noon Position			Day's Run	
	Latitude		Longitude	H.U.W.	
Fix				H.P.	
D.R.				True Dist. Run	
	Draft			True Av. Speed	
	Time	Fore	Aft	Dist. Run by Log	
Morning				Av. Speed by Log	
Evening				R.P.M.	
Departure & Arrival				Av. Speed	

Total Between Ports	From	To	Current Set & Drift	
	H.U.W.		Noon-Noon	
	H.P.			
	True Dist. Run		Point-Point	
	True Av. Speed			
	Dist. Run by Log		Tide	
	Av. Speed by Log			
	R.P.M.		Anchor Bearing	
	Av. Speed			

Wind Force（ビューフォート風力階級）

0	Calm	0 ～ 1.0	kn't未満
1	Light air	1.0 ～ 4.0	kn't未満
2	Light br'ze	4.0 ～ 7.0	kn't未満
3	Gentle br'ze	7.0 ～ 11.0	kn't未満
4	Mod. br'ze	11.0 ～ 17.0	kn't未満
5	Fresh br'ze	17.0 ～ 22.0	kn't未満
6	Strong br'ze	22.0 ～ 28.0	kn't未満
7	Near gale	28.0 ～ 34.0	kn't未満
8	Gale	34.0 ～ 41.0	kn't未満
9	Strong gale	41.0 ～ 48.0	kn't未満
10	Storm	48.0 ～ 56.0	kn't未満
11	Violent storm	56.0 ～ 64.0	kn't未満
12	Hurricane	64.0	kn't以上

Wave Force（風浪階級）

0	Calm	鏡のよう
1	Very smooth	さざなみ
2	Smooth	小波
3	Slight	やや波がある
4	Moderate	かなり波がある
5	Rought	波がやや高い
6	Very rought	波がかなり高い
7	High	相当荒れている
8	Very high	非常に荒れている
9	Phenomenal	異常な状態

Weather（天気）

b	Blue sky	快晴
bc	Fine but cloudy	晴れ
c	Cloudy	曇り
o	Over cast	全曇
r	Rainy	雨
q	Squalls	スコール
s	Snow	雪
f	Foggy	霧

4

4.6 Engineer's Log Book Writing E

① **Read the following Engineer's log book.**

② **Put mark in the Engineer's log book.**

③ **Calculate value and write down in Tables 1 and 2.**

10th Oct. 2001 <Left from Yuge>

1200 S/B eng'.
1205 Tried eng' and find it in good condition.
1209 Let go all shore lines and left Yuge.
1212 Slow ah'd eng'.
1215 Stop'ed eng'. <65435670>
1218 Slow ast' eng'.
1220 Full ast' eng'.
1222 Half ast' eng'.
1223 Stop'ed eng'. <65438900>
1224 Slow ah'd eng'.
1228 Half ah'd eng'.
1230 R/up eng'.

11th Oct. 2001 <Arrived at Miyazaki>

1100 S/B eng'.
1105 Slow down eng'.
1127 Stop'ed eng'. <65894690>
1130 Slow ah'd eng'.
1133 Stop'ed eng'. <658XXXX>
1138 Slow ah'd eng'.
1140 Full ast' eng'.
1145 Stop'ed eng'. <65897450>
1151 Sent first shore line and arrived at Miyazaki.

Standard Phrases （定例文）

Let go all shore lines	係留索を離す
Heave up anchor	錨を上げる
Let go anchor	錨を下ろす
Stop'ed eng'. <12345670>	機関停止
Left Takamatsu	高松出港
R/up eng'	機関用意解除
S/B eng'	機関用意
Arrived at Fukuyama	福山入港
Sent first shore line	係留索を渡す

Engine Orders（機関使用命令）

Tried eng'	機関試運転
Dead slow ah'd eng'	最微速前進
Dead slow ast' eng'	最微速後進
Slow ah'd eng'	微速前進
Slow ast' eng'	微速後進
Half ah'd eng'	前進半速
Half ast' eng'	後進半速
Full ah'd eng'	全速前進
Full ast' eng'	全速後進
Emergency full ast' eng'	緊急後進
Finished with engine	機関終了

〔Table 1〕 10th Oct. 2001

	Value	Remarks
H.U.W.		
H.P.		
H.E.M.		
Total Rev.		
R.P.M.		
Prop' Dist.		
LOG Dist.		
O.G. Dist.		
Slip		

〔Table 2〕 11th Oct. 2001

	Value	Remarks
H.U.W.		
H.P.		
H.E.M.		
Total Rev.		
R.P.M.		
Prop' Dist.		
LOG Dist.		
O.G. Dist.		
Slip		

H.U.W. (Hours Under Weigh)	航海時間
H.P. (Hours Propelling)	航進時間
H.E.M. (Hours Engine Motion)	機関使用時間
Average speed	平均速力
Total rev. (Revolutions)	積算回転数
R.P.M.	平均回転数
Hours temporary anchorage/drifting	仮泊時間
Hours in port	停泊時間
Prop' Dist. (Distance)	プロペラ航走距離
LOG Dist. (Distance)	対水航走距離
O.G. Dist. (Distance)	対地航走距離
Slip	スリップ

4.7 Basic Safety

N E

Writing. Give the definition and use of the following equipment, apparatus or signal.

① Life ring

② Life jackets

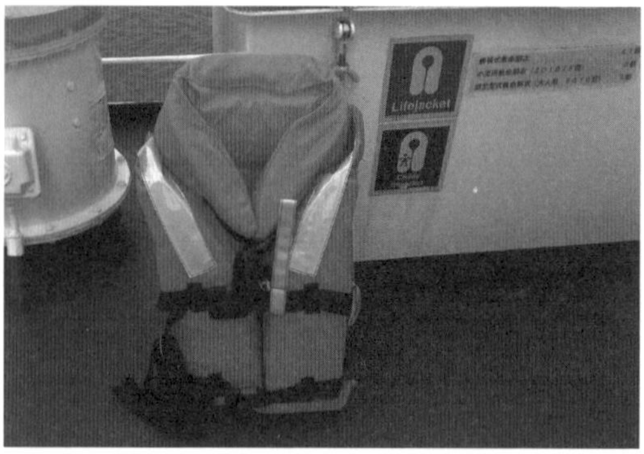

③ CO_2 fire extinguishing equipment

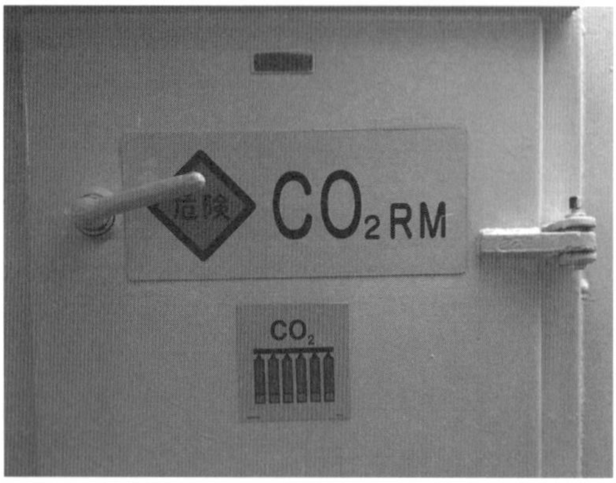

④ Cold water immersion suit

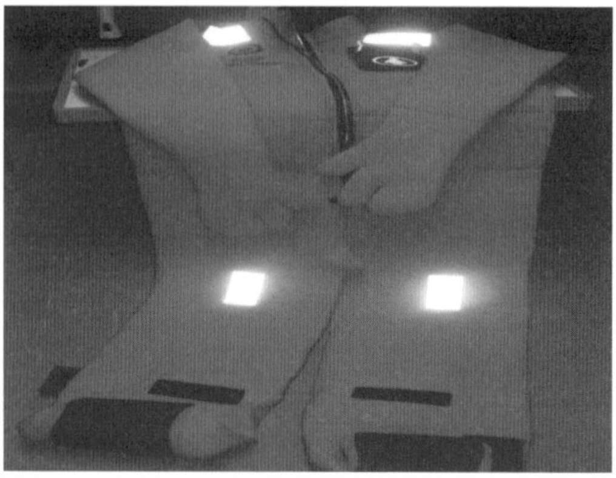

⑤ Fire extinguisher

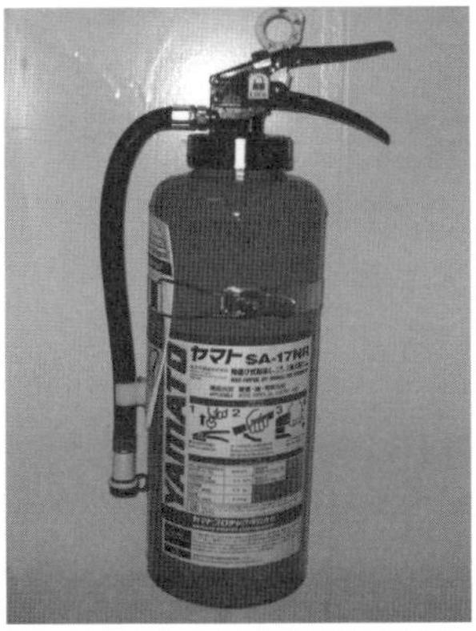

⑥ Fire station

⑦ Muster list

⑧ Escape trunk (route)

⑨ Life saving plan

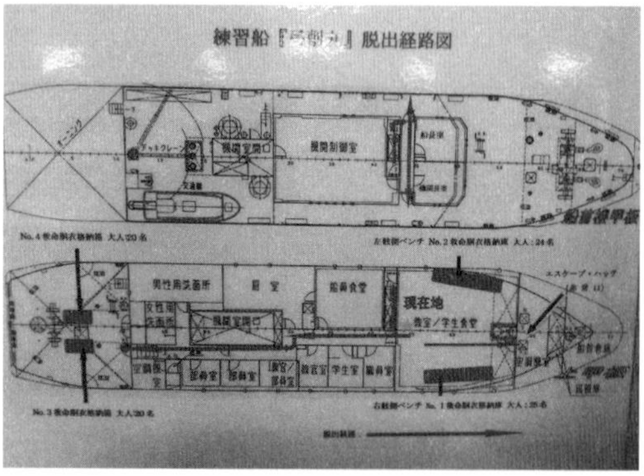

⑩ General arrangement plan

⑪ Muster station (assembly area)

⑫ Signages

⑬ Fire alarm panel

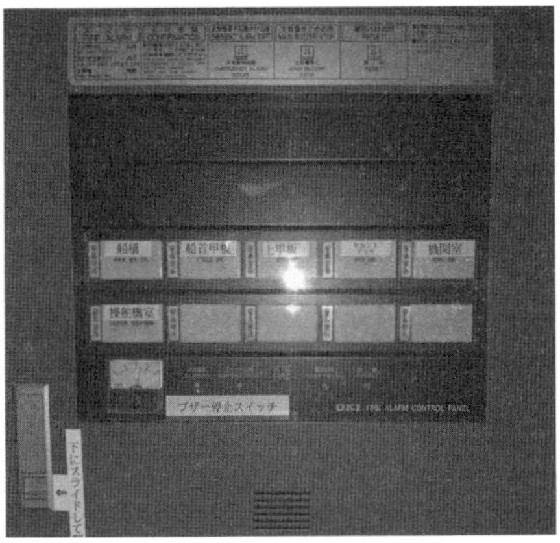

⑭ Signages

⑮ Lifeboats

⑯ Inflatable liferaft

4

⑰ Fire alarm button

⑱ Alarm bell

⑲ Smoke detector

⑳ Fire hydrant (on deck)

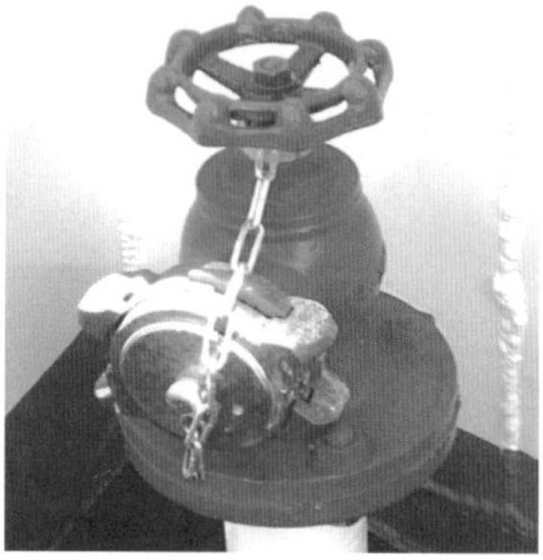

㉑ Signages

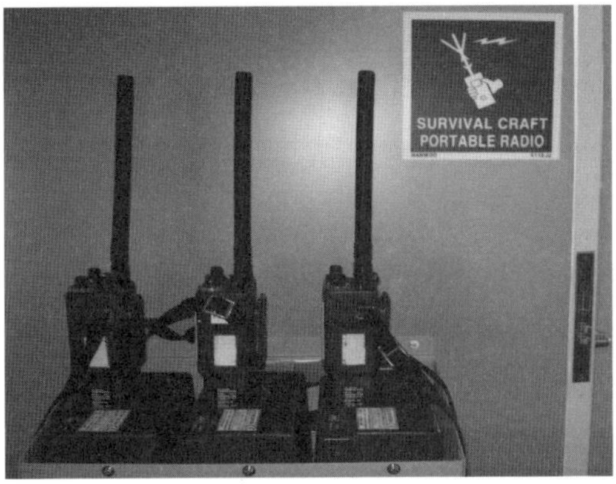

㉒ Signages

㉓ Lifebuoy

㉔ Liferaft instruction

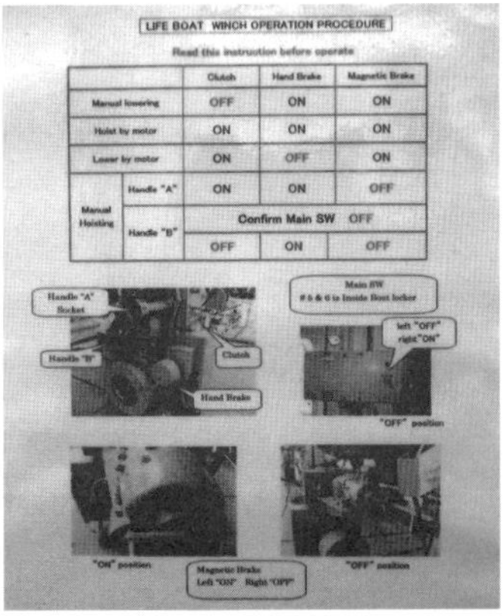

LIFE BOAT WINCH OPERATION PROCEDURE

Read this instruction before operate

		Clutch	Hand Brake	Magnetic Brake
Manual lowering		OFF	ON	ON
Hoist by motor		ON	ON	ON
Lower by motor		ON	OFF	ON
Manual Hoisting	Handle "A"	ON	ON	OFF
	Handle "B"	Confirm Main SW OFF		
		OFF	ON	OFF

Answers

1.1
1 ①6 ②9 ③1 ④5 ⑤7 ⑥11 ⑦10 ⑧2 ⑨8 ⑩13 ⑪14 ⑫4 ⑬3 ⑭12
2 ① Bow ② Anchor ③ After deck ④ Ensign ⑤ Hawsehole ⑥ Railings ⑦ Keel ⑧ Bulwark

1.2
① Helmet ② Gloves ③ Coverall ④ Safety shoes ⑤ Lifeboat ⑥ Immersion suit ⑦ Goggles ⑧ Earmuff ⑨ Face mask ⑩ Harness

1.3
① c ② b ③ b ④ a ⑤ c ⑥ b ⑦ c ⑧ b

1.4
1 ①6 ②2 ③7 ④11 ⑤3 ⑥8 ⑦1 ⑧5 ⑨9 ⑩4 ⑪10 ⑫12 ⑬13
2 ① Telegraph ② Magnetic compass ③ AIS ④ Gyro compass ⑤ Telephone ⑥ Wing angle meter ⑦ VHF radio

1.5
1 ①3 ②8 ③5 ④10 ⑤2 ⑥6 ⑦9 ⑧1 ⑨4 ⑩7
2 ① Clutch ② Motor ③ Frame ④ Reduction gear ⑤ Drum

1.6
Anchoring
→ Starbord anchor → Port anchor
→ Shackle → Holding power
→ Shank → Arm, Crown pin, Anchor head, Anchor ring
→ Stockless anchor → Stock anchor

1.7
Weather broadcasting
→ Pressure → High pressure, Cylone, Tropical storm, Typhoon
→ Cloud → Cirrus, Cirrocumulus, Altocumulus, Altostratus
→ Bureau → The Meteorological Agency, Meteorological observatory
→ Weather → Clear sky, Cloud, Rain, Snow, Thunder, Tornado

1.8
① channel ② great circle ③ latitude ④ longitude ⑤ compass ⑥ deviation ⑦ bearing ⑧ nautical ⑨ mile ⑩ life jacket ⑪ zenith ⑫ course ⑬ magnetic ⑭ gyro ⑮ axis ⑯ north ⑰ passage ⑱ speed ⑲ track ⑳ fairway ㉑ protection ㉒ mercator ㉓ waypoint ㉔ navigation ㉕ quarter

1.9
① Lateral ② Cardinal mark ③ Channel ④ Seaward ⑤ Obstruction ⑥ Buoy ⑦ Conical ⑧ Navigable ⑨ Waterline ⑩ Pillar ⑪ Direction ⑫ Starboard ⑬ Datum ⑭ Portside ⑮ Spar

1.10
① The VTS asks information from the ship.
② The vessel traffic services is established to improve the safety of the ships.
③ The VTS ordered the ships to observe the fairway speed.
④ When the ship arrives to the reporting point, she has to report to the VTS.
⑤ Observing the rules in the traffic separation scheme adds.

1.11

E	N	I	A	R	N	A	M	O	V	E	M	E	N	T
V	I	S	I	B	I	L	I	T	Y	C	J	M	O	R
P	R	E	S	S	U	R	E	A	G	R	A	N	I	O
S	U	I	D	A	R	L	L	O	N	O	K	D	T	P
L	S	E	L	I	M	E	T	E	R	F	B	I	I	I
A	W	I	N	D	I	M	I	L	E	B	A	R	S	C
C	E	V	I	N	S	P	E	E	D	E	C	A	O	A
S	L	G	R	F	T	N	A	C	I	A	K	A	P	L
A	L	P	L	O	O	V	O	E	M	U	I	E	L	A
P	U	T	C	L	O	G	R	W	D	F	N	I	E	O
O	N	J	C	E	A	N	S	U	S	O	G	S	N	I
T	A	Y	M	L	E	I	V	E	E	R	I	N	G	T
C	C	G	E	L	O	C	R	Y	T	T	O	G	O	U
E	D	F	O	R	E	I	M	R	O	T	S	V	E	M
H	Y	D	R	O	L	O	G	I	C	A	L	R	J	N

1.12 1 ①1 ②2 ③5 ④4 ⑤3 ⑥6 ⑦11 ⑧12 ⑨7 ⑩8 ⑪9 ⑫10

2 ① Target ② DCPA ③ TCPA ④ Track ⑤ Band

1.13 ① North ② South ③ East ④ West ⑤ Danger ⑥ Safe water ⑦ Starboard ⑧ Port side ⑨ Special

1.14 1 ①8 ②4 ③11 ④14 ⑤6 ⑥10 ⑦2 ⑧5 ⑨9 ⑩12 ⑪3 ⑫13 ⑬1 ⑭7

2 ① Flywheel ② Governor ③ Reduction gear ④ Turbocharger ⑤ Turning gear ⑥ Intermediate shaft ⑦ Bet

1.15 Main engine
- →Air → Supercharger, Exhaust turbocharger
- →Pump → FO (Fuel Oil), LO (Lubricating Oil), CFW (Cooling Fresh Water), CSW (Cooling Sea Water)
- →Valve → FO (Fuel Oil), CSW (Cooling Sea Water), CFW (Cooling Fresh Water), LO (Lubricating Oil)
- →Cylinder head → Intake valve, Exhaust valve, Fuel injection valve, Safety valve, Indicator valve

1.16 1 ①13 ②4 ③9 ④1 ⑤15 ⑥6 ⑦2 ⑧7 ⑨12 ⑩10 ⑪14 ⑫3 ⑬11 ⑭5 ⑮8

2 ① Voltmeter ② Amperemeter ③ Watt meter ④ Frequency meter ⑤ Governor ⑥ BUS ⑦ No fuse breaker ⑧ Synchro scope

1.17 Electric generator
- → Meter → Voltage, Current, Electric power, Frequency, Power factor
- → Parallel running → ACB (Air Circuit Breaker), Synchroscope, Governor, Load
- → BUS → Earth, Transformer, Battery, MSB (Main Switch Board)

1.18 ① duty ② main engine ③ standing order ④ bellbook ⑤ machinery ⑥ temperature ⑦ speed ⑧ control ⑨ propeller ⑩ cylinder ⑪ suction ⑫ shaft ⑬ inlet valve ⑭ economizer ⑮ generator ⑯ revolution ⑰ lube oil ⑱ pump ⑲ fuel ⑳ bunker

1.19 ① Generator ② Breaker ③ Insulator ④ Fuse ⑤ Circuits ⑥ Load ⑦ Motor ⑧ Volts ⑨ Watt ⑩ Current ⑪ Relay ⑫ Ampere ⑬ Voltage ⑭ Running ⑮ Electronic ⑯ Switch ⑰ Solenoid ⑱ Switchgear ⑲ Status lamp ⑳ Standby ㉑ Transformer ㉒ Control

1.20 1 ① 11 ② 1 ③ 8 ④ 13 ⑤ 4 ⑥ 7 ⑦ 2 ⑧ 10 ⑨ 5 ⑩ 9 ⑪ 12 ⑫ 6 ⑬ 3
2 ① Crank ② Cam ③ Piston ring ④ Manifold ⑤ Rod ⑥ Cylinder ⑦ Exhaust valve

1.21 1 ① 6 ② 10 ③ 2 ④ 8 ⑤ 7 ⑥ 3 ⑦ 9 ⑧ 1 ⑨ 11 ⑩ 4 ⑪ 5
2 ① Magnetic field ② Bearing bracket ③ Electric current ④ Torque ⑤ Coil

1.22 ① Delivery ② Impellers ③ Shaft ④ Suction ⑤ Air vent valve ⑥ Coupling ⑦ Bearing case ⑧ Mechanical seal ⑨ Shaft key ⑩ Volume case

1.23 ① plate ② three ③ fresh water ④ fresh water ⑤ steam ⑥ seawater ⑦ fuel oil ⑧ electric

1.24 ① 3 ② 4 ③ 1 ④ 5 ⑤ 2 ⑥ 7 ⑦ 8 ⑧ 6

1.25 ① Motor ② Compressor belt cover ③ Air compressor cylinder ④ suction discharge valve ⑤ Intake port ⑥ Reload/unreload ⑦ Safety valve ⑧ Discharge line

1.26 ① Impeller ② Distributor ③ Disc ④ Bowl nut ⑤ Bowl hood ⑥ Bowl body ⑦ Main cylinder

1.27 ① b ② c ③ a

1.28

S	U	P	E	R	C	H	A	R	G	E	R	J	D	V
Y	T	I	S	N	E	D	N	O	I	T	I	N	G	I
V	A	P	O	R	C	O	R	R	O	S	I	O	N	S
K	N	A	T	N	O	I	S	S	E	R	P	M	O	C
E	N	G	I	N	E	M	A	C	H	E	A	T	X	O
N	D	B	L	O	W	B	Y	O	A	C	I	L	I	S
E	I	P	I	P	E	S	E	M	B	G	A	E	D	I
T	E	L	L	O	U	V	N	B	R	R	E	N	E	T
L	S	J	A	R	I	N	R	U	A	E	M	I	D	Y
A	E	I	V	T	V	Y	U	S	S	K	A	L	I	F
H	L	E	I	O	G	O	F	T	I	N	R	A	R	U
P	Y	D	A	C	I	L	L	I	V	U	I	C	I	E
S	D	A	F	U	E	L	U	O	E	B	N	L	J	L
A	L	U	M	I	N	A	S	N	A	H	E	A	V	Y

1.29 ① valves ② body ③ brass ④ bonnet ⑤ cast iron ⑥ stainless steel ⑦ bow screw ⑧ gland ⑨ disc ⑩ valve seat ⑪ stem ⑫ sleeve ⑬ packing ⑭ actuator ⑮ globe valve ⑯ gate valve

1.30 ① Water level system ② Logic controller ③ Safety system ④ Control system ⑤ Burner

1.31 ① (2) status ② (3) feed water ③ (3) fuel oil ④ (4) breaker ⑤ (5) as necessary ⑥ (2) in ⑦ (3) burner's ⑧ (4) pressure ⑨ (2) flame propagation ⑩ (2) boiler's

2.2 1 ① Instruction ② Request ③ Warning ④ Question ⑤ Advice ⑥ Instruction ⑦ Intention ⑧ Answer

2 ① C ② H ③ A ④ E ⑤ G ⑥ F ⑦ D ⑧ B

3.6 Set A

① c ② c ③ a ④ b ⑤ a ⑥ b ⑦ c ⑧ a

Set B

① b ② c ③ a ④ a ⑤ c ⑥ b ⑦ b ⑧ a

Set C

① b ② a ③ b ④ c ⑤ c ⑥ c ⑦ c ⑧ a

4.1 Picture 1

- Unstrapped helmet might fall off and do not protect his head.
- Screw driver on his chest pocket will hurt him.
- Untied shoelace will make him slip.

Picture 2

・Hand of A be hit by hammer of B (extra space for second wrench).

Picture 3

・Wrench of B will fall and injure C.

・Tools must be in a tool box and just hoisted for hands to hold rails.

・Climb one at a time.

Picture 4

・Did not use face shield.

・Electrical wire coils around his body.

・Welding near a fuel valve might cause fire.

Picture 5

・A might fall because of not wearing harness.

・A will injure his spine due to the weight of the pulley with the chain.

Picture 6

・A and B do not wear helmet as the belt might snap/break and hit them.

・They do not put off electrical power while working on the compressor.

Picture 7

・A is not wearing a helmet instead just a bull cap.

・A is stepping on the mooring line (foot might be coiled).

Picture 8

・B should not be standing in front of A (he might be hit with the throwing line).

Picture 9

・Wrong size of the net for the used dunnage as load is not properly secured and might fall as it will hit the crew.

Picture 10

・A and B will injure their back/spine as the box is too heavy.

Picture 11

・B could lose his balance and fall into the sea.

・Because of the ship is rolling, A and B could fall into the sea and be drowned.

Picture 12

・As it is raining, he could get an electric shock when his screwdriver becomes wet.

・He could stumble over tools and electric wire and fall down.

Picture 13

・Because he is walking across a stepladder holding heavy parcels in both his hands, he could slip and fall into the sea.

・The stepladder could move and he could fall into the sea.

・The stepladder could break, and he could fall into the sea.

Picture 14

・As he is working standing on a chair, he could lose his balance and fall.

・Suddenly the clock will be off, it could hit the person below.

Picture 15

・The packages that C is shouldering could drop behind him and hit and injure D.

・As the men are not keeping enough space between them, if A fall backward, others will also fall.

・As many people are carrying packages, the gangway will bend under their weight and bounce up and down, so that they could lose their balance, drop their packages and injure someone.

Picture 16

・Because of the ship's rolling, the bosun's chair will swing wildly and he could be injured.

・As both his hands are occupied he will fall, if he loses his balance and his grip on the rope.

Picture 17

・A will trip over B's foot and fall.

・If A hits B, B could fall into the sea.

・As B leans over the handrail, he could lose his balance when the ship rolls, and fall.

・B could drop the paint can by accident, and hurt someone.

4.2 ① Put sign in the bridge that somebody is working on the radar.
② Wear harness while working aloft.
③ Stay away from line in order to avoid hit by cutting line.

4.3 ① Rope off the painted area.
② Wear gloves and goggles.
③ Wear goggles.

4.4 Story 1
[1] ① Fire in the main deck ② Report fire to the navigation bridge
③ Officer of the watch and captain ④ Muster station ⑤ Fight the fire

Story 2
[1] ① He lost his ticket at the airport ② At the airport ③ Osaka Japan ④ Reprinted the ticket
[2] ① B ② A ③ B

Story 3
[1] ① Smart phone ② A discount on the unit ③ The manager gives discount
④ 20% discount

Story 4
[1] ① B ② B ③ A

Story 5
[1] ① In the restaurant ② Spicy Ramen ③ There are insects in the food ④ No

4.7 ① Ring shape floating device to evacuate from ship at an emergency.
② Jacket shape floating device to evacuate from ship at an emergency.
③ Fire extinguishing facility with carbon dioxide gas.
④ Immersion suite to be able to resist cold sea water at evacuation.
⑤ Fire extinguishing device with chemical powder.
⑥ Fire fighting facility installed with water faucet and hose.
⑦ List to show casting of organization at an emergency.

⑧ Sign to show escape route at an emergency.
⑨ Ship route map which shows evacuation route at an emergency.
⑩ General arrangement plan to show ship overview.
⑪ Sign to show meeting point at evacuation.
⑫ Signages to show "no fires" and "off limit".
⑬ Panel to receive signal from fire alarm button and smoke detector.
⑭ Signages to show instruction of fire extinguisher.
⑮ Small boat to evacuate from ship at an emergency.
⑯ Inflatable raft to evacuate from ship at an emergency.
⑰ Button to report fire and other emergency.
⑱ Bell to warn fire and other emergency.
⑲ Device to tell if there is smoke.
⑳ Water faucet for fire fighting.
㉑ Signages to show the home position of portable VHF radio.
㉒ Signages to show the home position of radar transponder.
㉓ Lifebuoy with seawater-activated lights and smoke candle to help discovery at night and day.
㉔ Instruction sign to show how to use liferaft.

Maritime English Vocabulary Terms

Navigation（航海）

Ship data（船舶概要）

Call sign	船名符字
By the head	おもて脚
By the stem	とも脚
Dead weight tonnage	最大積載重量トン数
Draught	喫水
Free board mark	満載喫水線標
Gross tonnage	総トン数
Hogging	中央が船首部･船尾部より上昇している状態
LBP (Length Between Perpendicular)	垂線間長
LOA (Length Over All)	全長
Net tonnage	純トン数
Sagging	中央が船首部･船尾部より沈下している状態
Trim	トリム

Hull structure（船体構造）

Ballast	バラスト
Bulkhead	隔壁
Keel	キール

Deck（甲板）

Aft station	船尾配置
Anchor	錨
Anchor cable	錨鎖
Awning	甲板上の天幕
Bit	単係柱
Bollard	双係柱
Bow	船首
Brake handle	制動ハンドル
Bridge deck	船橋甲板
Bulwark	防波壁，舷墻
Cable stopper	制鎖器
Capstan	縦軸ウインチ，キャプスタン
Cat walk	狭通路
Chain cable compressor	制鎖器
Cleat	索留め
Clutch lever	クラッチレバー
Control box	操作箱
Escape hatch	緊急避難口
Fair leader	導索器
Flag hook	旗留金
Flag line	旗索
Flag staff	旗竿
Flying bridge	最上船橋
Forecastle deck	船首楼甲板
Forward station	船首配置
Freeboard	乾舷
Funnel	煙突
Gangway	舷門
Hand rail	手摺り
Hawse pipe	錨鎖孔
Hawser reel	錨鎖リール
Hydraulic crane	揚貨機
Mooring hole	係船孔
Mooring winch	係船機
Navigation bridge	航海船橋
Navigation lights	航海灯
Out-side passage	外部通路
Poop deck	船尾甲板
Port	左舷
Port side light (Red light)	左舷灯（赤色灯）
Propeller	推進機
Railing	手摺り
Recess hatch	リセルハッチ
Scuttle	小窓
Shell plate	外板
Sky light	天窓
Standing roller	スタンディングローラー
Starboard	右舷
Starboard side light (Green light)	右舷灯（緑色灯）
Stern	船尾
Stern light	船尾灯
Trap	舷梯
Upper deck	上甲板
Ventilater	通風筒
Warping end	綱巻胴
Wharf ladder	ワーフラダー

Wheel house	操舵室
Wind breaker	防風
Windlass	揚錨機

Bridge（船橋）	
AIS (Automatic Identification System)	自動船舶識別装置
Anemometer	風速計
Anemoscope	風向計
Chronometer	船内時計
ECDIS (Electronic Chart Display and Information System)	電子海図情報表示システム
Echo sounder	音響測深機
GPS (Global Positioning System)	衛星測位システム
Ground speed meter	対地速力計
Gyro compass	回転羅針儀
Log speed meter	対水速力計
LORAN (Long-Range Navigation)	ロラン
Magnetic compass	磁気羅針儀
Mike	マイク
Radar	電波探知機
Rudder angle meter	舵角計
Telegraph	電信機
Telephone	電話
Thruster controller	スラスター制御装置
VHS radio	VHSラジオ
Wheel	舵輪
Wing angle meter	翼角計

Wheel orders（操舵号令）	
Hard a port	左舵一杯
Hard a starboard	右舵一杯
Midship	中央
Port	左舵
Starboard	右舵
Steady	当舵

Chart work（海図）	
Bay	湾
Berth	埠頭･桟橋の係留場所
Buoy	浮標
Cape	岬
Channel	海峡
Latitude	緯度
Lighthouse	灯台
Longitude	経度
Nautical mile	海里
Peninsula	半島
Water depth	水深

Port radio（海岸局）	
MARTIS (Marine Traffic Information Service)	海上交通センター

Meteorological chart（天気図）	
Celsius degrees	摂氏
Clear	快晴
Cold front	寒冷前線
Cumulonimbus	積乱雲
Cumulus	積雲
Fahrenheit degrees	華氏
High pressure	高気圧
Humidity	湿度
Low pressure	低気圧
Meteorological observatory	気象台
Precipitation	降水量
Pressure pattern	気圧配置
Pressure ridge	気圧の尾根
Pressure trough	気圧の谷
Stratocumulus	層積雲
Stratus	層雲
Swell	うねり
Thunderstorm	雷を伴った激しい雨
Typhoon	台風
Warm front	温暖前線
Wind direction	風向き

Engine (機関)

General (全般)

AUTO (Automatic)	自動
Aux (Auxiliary)	補助
Charge	満たす
Close	閉める
C/R (Control Room)	機関制御室
Decrease	減少
Delivery	吐出
Discharge	放出
Emergency	非常
E/R (Engine Room)	機関室
Exhaust	排気
Flow meter	流量計
Forward	前進
IHP (Indicated Horse Power)	図示馬力
IN (Inlet)	入口
Increase	増加
Level	水位
Load	負荷
Local	現場
Main	主
MAN (Manual)	手動
Open	開く
OUT (Outlet)	出口
PRESS (Pressure)	圧力
Pressure gauge	圧力計
Remote	遠隔
Reverse	後進
Revolution	回転数
Revolution counter	積算回転計
RPM (Revolution Per Minute)	毎分回転数
Scavenging	掃気
SHP (Shaft Horse Power)	軸馬力
Start	始動
Stop	停止
Suction	吸入
Tachometer	速度計
TEMP (Temperature)	温度
Thermometer	温度計
Thrust	推力
Torque	回転力
Viscosity	粘度
Water gauge	水位計

Fluid (流体)

Bilge	油水混合物
CFW (Cooling Fresh Water)	冷却清水
Control Air	制御空気
CSW (Cooling Sea Water)	冷却海水
DW (Drinking Water)	飲料水
FO (Fuel Oil)	燃料油
FW (Fresh Water)	清水
LO (Lubricating Oil)	潤滑油
Sanitary	衛生の
Service Air	雑用空気
Steam	蒸気
SW (Sea Water)	海水

Electric (電気)

A.C. (Alternating Current)	交流
ACB (Air Circuit Breaker)	気中遮断器
Battery	蓄電池
BUS	母線
Current	電流
D.C. (Direct Current)	直流
D/G (Diesel Generator)	ディーゼル発電機
Earth	接地
Electric power	電力
FP (Feeder Panel)	給電盤
Frequency	周波数
Frequency meter	周波数計
GCP (Group Control Panel)	集合制御盤
Generator	発電機
GSP (Group Starter Panel)	集合始動盤
Motor	原動機
MSB (Main Switch Board)	主配電盤

NFB (No Fuse Breaker)	ヒューズなし遮断器
OC (Over Current)	過電流
Parallel running	並列運転
Power factor	力率
SC (Shore Connection)	陸上電源
Source	電源
Starter	始動器
Transformer	変圧器
Volt meter	電圧計
Voltage	電圧
Watt meter	電力計

Machinery etc.（機器）	
Air compressor	空気圧縮器
Air reservoir	空気槽
Bow thruster	船首スラスター
BRG (Bearing)	軸受
Capstan	縦軸ウインチ
Circulation PP	循環ポンプ
CLR (Cooler)	冷却器
Condenser	凝縮器
CPP (Controllable Pitch Propeller)	可変ピッチプロペラ
CYL (Cylinder)	気筒
Defrost heater	除霜器
ERC (Engine Remote Control)	機関の遠隔操縦
Evaporator	蒸発器
EXH GAS (Exhaust Gas)	排気ガス
EXH T/C (Exhaust Turbocharger)	排気ターボ
Expansion TK	膨張タンク
Filter	濾し器
Funnel	煙突
G/E (Generator Engine)	原動機
Gland	パッキン押さえ
Governor	調速機
GS PP (General Service pump)	雑用ポンプ
HTR (Heater)	加熱器
Ice chamber	冷凍庫
Incinerator	焼却炉
Intermediate shaft	中間軸
M/E (Main Engine)	主機関
Oily water separator	油水分離機
OVBD.V (Overboard Discharge Valve)	船外弁
Plumber block	中間軸受
PP (Pump)	ポンプ
Pressure TK	圧力タンク
Purifier	清浄機
Refrigerator	冷蔵庫
Reserve TK	予備タンク
Rocker arm	揺れ腕
Service TK	常用タンク
Settling TK	澄ましタンク
Shift PP	移送ポンプ
Side thruster	サイドスラスター
Steering machine	操舵機
Stern thruster	船尾スラスター
Stern tube	船尾管
Storage TK	貯蔵タンク
Strainer	ろ過器
Supercharger	過給器
TK (Tank)	タンク
Ventilator	通風装置
Windlass	揚錨機

The Others（その他）

Ship positions（船舶職級）	
Captain	船長
First mate/officer	一等航海士
Second mate/officer	二等航海士
Third mate/officer	三等航海士
Quarter master	操舵手
Bosun/Boatswain	甲板長
Able bodied seaman/AB	熟練船員
Ordinary seaman	甲板員
Chief cook	司厨長
Chief engineer	機関長
First engineer	一等機関士
Second engineer	二等機関士
Third engineer	三等機関士
Forth engineer	四等機関士
Oiler	操機手
Wiper	操機員
Instructor	教官
Cadet	実習生
Pilot	水先人

Ship types（船種）	
Bulk ship	ばら積み船
Chip carrier	チップ専用船
Coal carrier	石炭運搬船
Container ship	コンテナ船
Ferry	フェリー
Fishing boat	漁船
Ore carrier	鉱石運搬船
Passenger ship	客船
Patrol ship	巡視船
Pure car carrier	自動車専用船
Refrigerated ship	冷凍運搬船
Roll-on/roll-off ship	ローロー船
Specialized vessel	特殊船
Anchor handling vessel	アンカーハンドリング船
Research ship	調査船
Supply vessel	供給船
Survey vessel	測量船
Tag	曳船，タグ
Tanker	タンカー
Chemical tanker	ケミカルタンカー
LNG (Liquefied Natural Gas) carrier	LNG船
LPG (Liquefied Petroleum Gas) carrier	LPG船
Oil tanker	石油タンカー
Product tanker	製品タンカー
Timber carrier	木材運搬船
Training ship	練習船

Basic safety（安全）	
Coverall	つなぎ
Earmuff	耳栓
Face mask	マスク
Gloves	手袋
Goggles	保護眼鏡
Harness	ハーネス
Helmet	ヘルメット
Safety shoes	安全靴

Life saving（救命）	
Escape trunk	避難経路
Fire alarm button	火災報知器
Fire extinguisher	消火器
Fire station	消火栓
Immersion suit	イマージョンスーツ
Life boat	救命艇
Life jacket	救命胴衣
Life raft	膨張式救命筏
Life ring	救命浮環
Muster station	非常時集合場所
Radar transponder	捜索救助用無線応答装置
Smoke detector	煙感知器

<編者紹介>

商船高専キャリア教育研究会

商船学科学生のより良きキャリアデザインを構想・研究することを目的に、2007年に結成。
富山・鳥羽・弓削・広島・大島の各商船高専に所属する教員有志が会員となって活動している。
2016年は富山高等専門学校が事務局を担当している。
連絡先：〒933-0293
　　　　富山県射水市海老江練合1-2
　　　　富山高等専門学校 商船学科 気付

ISBN978-4-303-23348-8

マリタイムカレッジシリーズ

Let's Enjoy Maritime English

2016年3月21日　初版発行

編　者　商船高専キャリア教育研究会　　検印省略
協　力　MAAP
発行者　岡田節夫
発行所　海文堂出版株式会社
　本社　東京都文京区水道2-5-4（〒112-0005）
　　　　電話 03(3815)3291(代)　FAX 03(3815)3953
　　　　http://www.kaibundo.jp/
　支社　神戸市中央区元町通3-5-10（〒650-0022）
日本書籍出版協会会員・工学書協会会員・自然科学書協会会員

PRINTED IN JAPAN　　印刷　田口整版／製本　誠製本